Autonomy Oriented Computing

From Problem Solving to Complex Systems Modeling

MULTIAGENT SYSTEMS, ARTIFICIAL SOCIETIES, AND SIMULATED ORGANIZATIONS
International Book Series

Series Editor: Gerhard Weiss, *Technische Universität München*

Editorial Board:
Kathleen M. Carley, Carnegie Mellon University, PA, USA
Yves Demazeau, CNRS Laboratoire LEIBNIZ, France
Ed Durfee, University of Michigan, USA
Les Gasser, University of Illinois at Urbana-Champaign, IL, USA
Nigel Gilbert, University of Surrey, United Kingdom
Michael Huhns, University of South Carolina, SC, USA
Nick Jennings, University of Southampton, UK
Victor Lesser, University of Massachusetts, MA, USA
Katia Sycara, Carnegie Mellon University, PA, USA
Michael Wooldridge, University of Liverpool, United Kingdom

Books in the Series:

CONFLICTING AGENTS: *Conflict Management in Multi-Agent Systems*, edited by Catherine Tessier, Laurent Chaudron and Heinz-Jürgen Müller, ISBN: 0-7923-7210-7

SOCIAL ORDER IN MULTIAGENT SYSTEMS, edited by Rosaria Conte and Chrysanthos Dellarocas, ISBN: 0-7923-7450-9

SOCIALLY INTELLIGENT AGENTS: *Creating Relationships with Computers and Robots*, edited by Kerstin Dautenhahn, Alan H. Bond, Lola Cañamero and Bruce Edmonds, ISBN: 1-4020-7057-8

CONCEPTUAL MODELLING OF MULTI-AGENT SYSTEMS: *The CoMoMAS Engineering Environment*, by Norbert Glaser, ISBN: 1-4020-7061-6

GAME THEORY AND DECISION THEORY IN AGENT-BASED SYSTEMS, edited by Simon Parsons, Piotr Gmytrasiewicz, Michael Wooldridge, ISBN: 1-4020-7115-9

REPUTATION IN ARTIFICIAL SOCIETIES: *Social Beliefs for Social Order*, by Rosaria Conte, Mario Paolucci, ISBN: 1-4020-7186-8

AGENT AUTONOMY, edited by Henry Hexmoor, Cristiano Castelfranchi, Rino Falcone, ISBN: 1-4020-7402-6

AGENT SUPPORTED COOPERATIVE WORK, edited by Yiming Ye, Elizabeth Churchill, ISBN: 1-4020-7404-2

DISTRIBUTED SENSOR NETWORKS, edited by Victor Lesser, Charles L. Ortiz, Jr., Milind Tambe, ISBN: 1-4020-7499-9

AN APPLICATION SCIENCE FOR MULTI-AGENT SYSTEMS, edited by Thomas A. Wagner, ISBN: 1-4020-7867-6

METHODOLOGIES AND SOFTWARE ENGINEERING FOR AGENT SYSTEMS: *The Agent-Oriented Software Engineering Handbook,* edited by Federico Bergenti, Marie-Pierre Gleizes, Franco Zambonelli

Autonomy Oriented Computing

From Problem Solving to Complex Systems Modeling

Jiming Liu
Xiaolong Jin
Kwok Ching Tsui
Hong Kong Baptist University

KLUWER ACADEMIC PUBLISHERS
BOSTON / DORDRECHT / LONDON

Library of Congress Cataloging-in-Publication Data

A C.I.P. Catalogue record for this book is available
from the Library of Congress.

Liu, Jiming
     Autonomy Oriented Computing: From Problem Solving to Complex Systems Modeling / by Jiming Liu, Xiaolong Jin, Kwok
Ching Tsui
     p.cm.

ISBN 978-1-4419-5480-0
e-ISBN 978-1-4020-8122-4   Printed on acid-free paper.

©2010 Kluwer Academic Publishers.

Printed in the United States of America.

9 8 7 6 5 4 3 2 1

springeronline.com

To my parents, my wife, Meilee,
and my two daughters, Isabella and Bernice,
who have given me life, love,
inspiration, and purpose.

Jiming Liu

To my wife, Zhen, and my parents,
for their continuous support and endless love.

Xiaolong Jin

To May and Abigail,
the source of unceasing love,
and God the Creator.

Kwok Ching Tsui

Contents

Contents

List of Figures

List of Tables

List of Algorithms

Preface:
Towards a New Computing Paradigm

With the advent of computing, we are fast entering a new era of discovery and opportunity. In business, market researchers will be able to predict the potential market share of a new product on-the-fly by synthesizing news reports, competitor analysis, and large-scale simulations of consumer behavior. In life and material sciences, specially engineered amorphous computational particles will be able to perform optimal search, whether they are bio-robot agents to kill cancer cells inside human bodies or smart paints to spread evenly over and fill cracks on rugged surfaces. In environmental sciences, surveillance applications will be able to deploy wireless, mobile sensor networks to monitor wild vegetation and route the tracking measurements of moving objects back to home stations efficiently and safely. In robotics, teams of rescue or Mars exploratory robots will be able to coordinate their manipulation tasks in order to collectively accomplish their missions, while making the best use of their capabilities and resources.

All the above examples exhibit a common characteristic, that is, the task of computing is seamlessly carried out in a variety of physical embodiments. There is no single multi-purpose or dedicated machine that can manage to accomplish a job of this nature. The key to success in such applications lies in a large-scale deployment of computational agents capable of autonomously making their localized decisions and achieving their collective goals.

We are now experiencing a world in which the traditional sense of computers is getting obsolete. It calls for a more powerful, intelligent computing paradigm for handling large-scale data exploration and information processing. We are in a critical moment to develop such a new computing paradigm in order to invent new technologies, to operate new business models, to discover new scientific laws, and even to better understand the universe in which we live.

In human civilizations, science and technology develop as a result of our curiosity to uncover such fundamental puzzles as who we are, how the universe evolves, and how nature works, multiplied by our desires to tackle such practical issues as how to overcome our limitations, how to make the best use of our resources, and how to sustain our well-being.

This book is a testimony of how we embrace new scientific and technological development in the world of computing. We specifically examine the metaphors of autonomy as offered by nature and identify their roles in addressing our practical computing needs. In so doing, we witness the emergence of a new computing paradigm, called autonomy oriented computing (AOC).

Autonomy Oriented Computing

While existing methods for modeling autonomy are successful to some extent, a generic model or framework for handling problems in complex systems, such as ecological, social, economical, mathematical, physical, and natural systems, effectively is still absent. Autonomy oriented computing (AOC) unifies the methodologies for effective analysis, modeling, and simulation of the characteristics of complex systems. In so doing, AOC offers a new computing paradigm that makes use of autonomous entities in solving computational problems and in modeling complex systems. This new paradigm can be classified and studied according to (1) how much human involvement is necessary and (2) how sophisticated a model of computational autonomy is, as follows:

AOC-by-fabrication: Earlier examples with this approach are entity-based image feature extraction, artificial creature animation, and ant colony optimization. Lifelike behavior and emergent intelligence are exhibited in such systems by means of fabricating and operating autonomous entities.

AOC-by-prototyping: This approach attempts to understand self-organized complex phenomena by modeling and simulating autonomous entities. Examples include studies on Web regularities based on self-adaptive information foraging entities.

AOC-by-self-discovery: This approach automatically fine-tunes the parameters of autonomous behaviors in solving and modeling certain problems. A typical example is using autonomous entities to adaptively solve a large-scale, distributed optimization problem in real time.

As compared to other paradigms, such as centralized computation and top-down systems modeling, AOC has been found to be extremely appealing in the following aspects:

- To capture the essence of autonomy in natural and artificial systems;

- To solve computationally hard problems, e.g., large-scale computation, distributed constraint satisfaction, and decentralized optimization, that are dynamically evolving and highly complex in terms of interaction and dimensionality;

- To characterize complex phenomena or emergent behavior in natural and artificial systems that involve a large number of self-organizing, interacting entities;

- To discover laws and mechanisms underlying complex phenomena or emergent behaviors.

Early Work on AOC

The ideas, formulations, and case studies that we introduce in this book have resulted largely from the research undertaken in the AOC Research Lab of Hong Kong Baptist University under the direction of Professor Jiming Liu. In what follows, we highlight some of the earlier activities in our journey towards the development of AOC as a new paradigm for computing.

Our first systematic study on AOC originated in 1996[1]. As originally referred to Autonomy Oriented Computation, the notion of AOC first appeared in the book of *Autonomous Agents and Multi-Agent Systems* (AAMAS)[2]. Later, as an effort to promote the AOC research, the First International Workshop on AOC was organized and held in Montreal in 2001[3].

Earlier projects at the AOC Lab have been trying to explore and demonstrate the effective use of AOC in a variety of domains, covering constraint satisfac-

---

[1] The very first reported formulation of cellular automaton for image feature extraction can be found in J. Liu, Y. Y. Tang, and Y. Cao. An Evolutionary Autonomous Agents Approach to Image Feature Extraction. *IEEE Transactions on Evolutionary Computation*, 1(2):141-158, 1997. J. Liu, H. Zhou, and Y. Y. Tang. Evolutionary Cellular Automata for Emergent Image Features. In Shun-ichi Amari et al., editors, *Progress in Neural Information Processing*, Springer, pages 458-463, 1996.

[2] J. Liu. *Autonomous Agents and Multi-Agent Systems: Explorations in Learning, Self-Organization, and Adaptive Computation*, World Scientific Publishing, 2001.

[3] At this workshop, a comprehensive introduction to this new research field, as the further development of AAMAS, was given; See J. Liu, K. C. Tsui, and J. Wu. Introducing Autonomy Oriented Computation (AOC). In *Proceedings of the First International Workshop on Autonomy Oriented Computation (AOC 2001)*, Montreal, May 29, 2001, pages 1-11.

tion problem solving[4], mathematical programming[5], optimization[6], image processing[7], and data mining[8]. Since 2000, projects have been launched to study the AOC approaches to characterizing (i.e., modeling and explaining) observed or desired regularities in real-world complex systems, e.g., self-organized Web regularities and HIV infection dynamics, as a white-box alternative to the traditional top-down or statistical modeling[9].

These AOC projects differ from traditional AI and agent studies in that here we pay special attention to the role of self-organization, a powerful methodology as demonstrated in nature and well suited to the problems that involve large-scale, distributed, locally interacting, and sometimes rational entities. This very emphasis on self-organization was also apparent in the earlier work on collective problem solving with a group of autonomous robots[10] and behavioral self-organization[11].

Recently, we have started to explore a new frontier, the AOC applications to the Internet. This work has dealt with the theories and techniques essential

[4]The first experiment that demonstrated the idea of cellular automaton-like computational entities in solving constraint satisfaction problems (CSP) can be found in J. Han, J. Liu, and Q. Cai. From ALife Agents to a Kingdom of N Queens. In J. Liu and N. Zhong, editors, *Intelligent Agent Technology: Systems, Methodologies, and Tools*, World Scientific Publishing, pages 110-120, 1999. Our recent work has extended the previous work by developing formal notions of computational complexity for AOC in distributed problem solving; See, X. Jin and J. Liu. Agent Networks: Topological and Clustering Characterization. In N. Zhong and J. Liu, editors, *Intelligent Technologies for Information Analysis*, Springer, pages 285-304, 2004.

[5]J. Liu and J. Yin. Multi-Agent Integer Programming. In *Lecture Notes in Computer Science*, Vol. 1983, Springer, pages 301-307, 2000.

[6]A successfully demonstrated application in optimization is to solve benchmark functional optimization problems with promising results; See, K. C. Tsui and J. Liu. Evolutionary Diffusion Optimization, Part I: Description of the Algorithm. In *Proceedings of the 2002 Congress on Evolutionary Computation (CEC 2002)*, Honolulu, Hawaii, May 12-17, 2002. K. C. Tsui and J. Liu. Evolutionary Diffusion Optimization, Part II: Performance Assessment. In *Proceedings of the 2002 Congress on Evolutionary Computation (CEC 2002)*, Honolulu, Hawaii, May 12-17, 2002.

[7]J. Liu and Y. Zhao. On Adaptive Agentlets for Distributed Divide-and-Conquer: A Dynamical Systems Approach. *IEEE Transactions on Systems, Man, and Cybernetics, Part A: Systems and Humans*, 32(2):214-227, 2002.

[8]J. Liu. *Autonomy Oriented Computing (AOC): A New Paradigm in Data Mining and Modeling*, Invited Talk, Workshop on Data Mining and Modeling, June 27-28, 2002, Hong Kong.

[9]The results were first reported in J. Liu and S. Zhang. Unveiling the Origins of Internet Use Patterns. In *Proceedings of INET 2001, The Internet Global Summit*, Stockholmsmssan, Stockholm, Sweden, June 5-8, 2001.

[10]J. Liu and J. Wu. *Multi-Agent Robotic Systems*, CRC Press, 2001. J. Liu and J. Wu. Evolutionary Group Robots for Collective World Modeling. In *Proceedings of the Third International Conference on Autonomous Agents (AGENTS'99)*, Seattle, WA, May 1-5, 1999. J. Liu. *Self-organization, Evolution, and Learning*, Invited Lectures by Leading Researchers, Pacific Rim International Workshop on Multi-Agents (PRIMA 2002) Summer School on Agents and Multi-Agent Systems, Aug. 17, 2002, Tokyo, Japan.

[11]J. Liu, H. Qin, Y. Y. Tang, and Y. Wu. Adaptation and Learning in Animated Creatures. In *Proceedings of the First International Conference on Autonomous Agents (AGENTS'97)*, Marina del Rey, California, Feb. 5-8, 1997. J. Liu and H. Qin. Behavioral Self-Organization in Synthetic Agents. *Autonomous Agents and Multi-Agent Systems*, Kluwer Academic Publishers, 5(4):397-428, 2002.

for the next paradigm shift in the World Wide Web, i.e., the Wisdom Web[12]. It covers a number of key Web Intelligence (WI) capabilities, such as (1) autonomous service planning; (2) distributed resource discovery and optimization[13]; (3) Problem Solver Markup Language (PSML); (4) social network evolution; (5) ubiquitous intelligence.

Overview of the Book

This book is intended to highlight the important theoretical and practical issues in AOC, with both methodologies and experimental cases studies.

It can serve as a comprehensive reference book for researchers, scientists, engineers, and professionals in the fields of computer science, autonomous systems, robotics, artificial life, biology, psychology, ecology, physics, business, economics, and complex adaptive systems, among others.

It can also be used as a text or supplementary book for graduate or undergraduate students in a broad range of disciplines, such as:

- Agent-Based Problem Solving;

- Amorphous Computing;

- Artificial Intelligence;

- Autonomous Agents and Multi-Agent Systems;

- Complex Adaptive Systems;

- Computational Biology;

- Computational Finance and Economics;

- Data Fusion and Exploration;

- Emergent Computation;

- Image Processing and Computer Vision;

- Intelligent Systems;

---

[12]J. Liu. Web Intelligence (WI): What Makes Wisdom Web? In *Proceedings of the Eighteenth International Joint Conference on Artificial Intelligence (IJCAI-03)*, Acapulco, Mexico, Aug. 9-15, 2003, pages 1596-1601, Morgan Kaufmann Publishers. J. Liu. *Web Intelligence (WI): Some Research Challenges*, Invited Talk, the Eighteenth International Joint Conference on Artificial Intelligence (IJCAI-03), Aug. 9-15, 2003, Acapulco, Mexico.

[13]One project addressed the issue of resource discovery and allocation; See, Y. Wang and J. Liu. Macroscopic Model of Agent Based Load Balancing on Grids. In *Proceedings of the Second International Joint Conference on Autonomous Agents and Multi-Agent Systems (AAMAS 2003)*, Melbourne, Australia, July 14-18, 2003. K. C. Tsui, J. Liu, and M. J. Kaiser. Self-Organized Load Balancing in Proxy Servers. *Journal of Intelligent Information Systems*, Kluwer Academic Publishers, 20(1):31-50, 2003.

- Modeling and Simulation;

- Nature Inspired Computation;

- Operations Research;

- Optimization;

- Programming Paradigms;

- Robotics and Automation;

- Self-Organization.

The book contains two parts. In Part I, Fundamentals, we describe the basic concepts, characteristics, and approaches of AOC. We further discuss the important design and engineering issues in developing an AOC system, and present a formal framework for AOC modeling. In Part II, AOC in Depth, we provide detailed methodologies and case studies on how to implement and evaluate AOC in problem solving (i.e., Chapter 5, AOC in Constraint Satisfaction and Chapter 7, AOC in Optimization) as well as in complex systems modeling (i.e., Chapter 6, AOC in Complex Systems Modeling). In these chapters, we start with introductory or survey sections on practical problems and applications that call for the respective AOC approach(es) and specific formulations. In Chapter 8, Challenges and Opportunities, we revisit the important ingredients in the AOC paradigm and outline some directions for future research and development.

The book contains numerous illustrative examples and experimental case studies. In addition, it also includes exercises at the end of each chapter. These materials further consolidate the theories and methodologies through:

- Solving, proving, or testing some specific issues and properties, which are mentioned in the chapter;

- Application of certain methodologies, formulations, and algorithms described in the chapter to tackle specific problems or scenarios;

- Development of new formulations and algorithms following the basic ideas and approaches presented;

- Comparative studies to empirically appreciate the differences between a specific AOC method or approach and other conventional ones;

- Philosophical and critical discussions;

- Survey of some literature and hence identification of AOC research problems in a new domain.

Moreover, we will make related electronic materials available on the Web for interested readers to download. These electronic materials include: source codes for some of the algorithms and case studies described in the book, presentation slides, new problems or exercises, and project demos. Details can be found at *http://www.comp.hkbu.edu.hk/~jiming/*.

Whether you are interested in applying the AOC techniques introduced here to solve your specific problems or you are keen on further research in this exciting field, we hope that you will find this thorough and unified treatment of AOC useful and insightful. Enjoy!

Hong Kong
Fall 2004

Jiming Liu
Xiaolong Jin
Kwok Ching Tsui

Acknowledgments

We wish to thank all the people who have participated in our AOC related research activities. In particular, we would like to express our gratitude to the following people who have studied or collaborated with us over the years as we embark on the journey to AOC: Yichuan Cao, Jing Han, Markus Kaiser, Oussama Khatib, Jun Le, Yunqi Lei, Hong Qin, Y. Y. Tang, Yi Tang, Yuanshi Wang, Jianbing Wu, Yiyu Yao, Yiming Ye, Shiwu Zhang, Yi Zhao, Ning Zhong, and Huijian Zhou.

We wish to thank the Computer Science Department at Hong Kong Baptist University for providing an ideal, free environment for us to develop this exciting research field.

We want to acknowledge the support of the following research grants: (1) Hong Kong Research Grant Council (RGC) Central Allocation Grant (HKBU 2/03/C) and Earmarked Research Grants (HKBU 2121/03E)(HKBU 2040/02E), (2) Hong Kong Baptist University Faculty Research Grants (FRG), (3) National Grand Fundamental Research 973 Program of China (2003CB316901), and (4) Beijing Municipal Key Laboratory of Multimedia and Intelligent Software Technology (KP0705200379).

We are grateful to Ms. Melissa Fearon, Computer Science Editor at Springer, and Dr. Gerhard Weiss, the series editor, for their encouragement and professional handling.

Most of all, we would like to offer our special thanks to our families and friends for their understanding, support, and love.

Other Credits

The theories and case studies as mentioned in some of the chapters in this book have been partially based on our earlier work:

- J. Liu, X. Jin, and K. C. Tsui. Autonomy Oriented Computing (AOC): Formulating Computational Systems with Autonomous Components. *IEEE Transactions on Systems, Man and Cybernetics, Part A: Systems and Humans* (in press).

- J. Liu, S. Zhang, and J. Yang. Characterizing Web Usage Regularities with Information Foraging Agents. *IEEE Transactions on Knowledge and Data Engineering*, 16(5):566-584, 2004.

- J. Liu, J. Han, and Y. Y. Tang. Multi-Agent Oriented Constraint Satisfaction *Artificial Intelligence*, 136(1):101-144, 2002.

- K. C. Tsui and J. Liu. Evolutionary Multi-Agent Diffusion Approach to Optimization. *International Journal of Pattern Recognition and Artificial Intelligence*, World Scientific Publishing, 16(6):715-733, 2002.

- J. Liu and Y. Y. Tang. Adaptive Segmentation with Distributed Behavior Based Agents. *IEEE Transactions on Pattern Analysis and Machine Intelligence*, 21(6):544-551, 1999.

Any omission of credit or acknowledgement will be corrected in the future editions of this book.

Hong Kong   *Jiming Liu*
Fall 2004   *Xiaolong Jin*
                                                               Kwok Ching Tsui

Part I
FUNDAMENTALS

Chapter 1

From Autonomy to AOC

1.1. Introduction

Autonomy oriented computing (AOC) is a new bottom-up paradigm for problem solving and complex systems modeling. In this book, our goal is to substantiate this very statement and to demonstrate useful AOC methodologies and applications. But, before we do so, we need to understand some of the most fundamental issues involved: What are the general characteristics of complex systems consisting of autonomous entities? What types of behavior can a single or a collection of autonomous entities exhibit or generate? How can we give a definition of autonomy based on the notion of behavior? In a bottom-up computing system, how can the property of autonomy be modeled and utilized? What types of problem is such a bottom-up computing paradigm indented to solve? How different is this AOC paradigm from other previous or current computing paradigms?

In this chapter, we will try to answer each of the above questions. These answers will provide a general context for our later discussions on the AOC formalisms, methodologies, and applications.

1.1.1 Complex Multi-Entity Systems

Examples of complex multi-entity systems are plentiful in everyday life. Traffic on motorways is notoriously busy but most drivers seem to have learned the type of skill to avoid almost all kinds of collision, with only few exceptions. Brokers in stock markets seem to have developed a highly sophisticated 'herding' behavior to buy and sell in the wake of market information. The balance between species of life forms in an ecosystem is equally complex and yet all of them seem to be settled into a dynamical equilibrium, most of the time. These scenarios point to a common phenomenon that can be observed in everyday life – many independent minds can sometimes maintain order in a global sense despite the lack of communication, central planning, or prior arrangement.

In contrast, team sports players, such as basketball players, spend many hours practicing team formations before they make 'magical' passes. Without such intensive practice, players will not be able to get the cue from others and defeat is imminent. Even when a team has been playing together for a long time, secret signs have to be given before desired results can be achieved. In the presence of a group of independent minds, team sports must be significantly different from motorway traffic so that different behavior results.

Nature is full of complex systems some of which have been extensively studied from different angles and with different objectives. Some researchers want to understand the working mechanism of a complex system concerned. Immunologists, for example, want to know the way in which the human immune system reacts to antigens [Louzoun et al., 2000]. Similarly, economists want to know the factors contributing to the ups and downs in share prices. The knowledge gained in this way helps scientists predict future systems behavior. Others studying complex systems behavior want to simulate the observed complex behavior and formulate problem solving strategies for hard computational problems, such as global optimization. Computer scientists and mathematicians have formulated various algorithms based on natural evolution to solve their problems at hand. In general, one wants to be able to explain, predict, reconstruct, and deploy a complex system.

1.1.2   Complex Systems Modeling

An important task common to the above studies is to build models of certain complex systems. Techniques for complex systems modeling can be broadly divided into top-down and bottom-up approaches. Top-down approaches start from the high-level characterization of a system and use various tools, such as ordinary differential equations. These approaches generally treat every part of a complex system homogeneously and tend to model average cases well, where the behavioral difference of the individuals is minimal and can be ignored [Casti, 1997]. However, this is not always applicable.

Bottom-up approaches, on the other hand, start with the smallest and simplest entities of a complex system and model their behavior as follows:

- **Autonomous:** System entities are rational individuals that act independently. In other words, a central controller for directing and coordinating individual entities is absent.

- **Emergent:** They exhibit complex behavior that is not present or predefined in the behavior of the autonomous entities.

- **Adaptive:** They often change their behavior in response to changes in the environment in which they are situated.

- **Self-organized:** They are able to organize themselves to achieve the above.

1.2. Basic Concepts and Taxonomies

Complex systems modeling using a bottom-up approach centers around the external behavior and internal behavior of individual entities. The trickiest part of the modeling task is to find the relationship between these two types of behavior. AOC adds a new dimension to the modeling process, i.e., modeling and deploying autonomy. Broadly speaking, autonomy is an attribute of entities in a complex system and autonomous entity is the building block of an AOC system. This section will first discuss different types of behavior and their relationships, and then define the notion of autonomy in the context of a computational system.

1.2.1 Types of Behavior

Entities in a complex system can perform certain primitive behavior as well as three types of complex behavior: emergent behavior, purposeful behavior, and emergent purposeful behavior.

Definition 1.1 (Primitive behavior) *The primitive behavior of an entity is the behavior that is governed by a set of predefined rules. These rules dictate how the states of the entity are updated. They are triggered by some internal or external stimuli.*

Definition 1.2 (Emergent behavior) *The emergent behavior of one or more entities is the behavior not inherent in the primitive behavior of an entity. It is achieved through nonlinear interactions among individual entities.*

It should be pointed out that emergent behavior may not be the same as collective behavior as it may not involve sharing of power or division of labor among individual entities.

Definition 1.3 (Purposeful behavior) *The purposeful behavior of one or more entities is the behavior that leads to certain desired states (i.e., goals) of entities.*

Definition 1.4 (Emergent purposeful behavior) *The emergent purposeful behavior of one or more entities is the emergent behavior that directs entities towards certain goals.*

It should be pointed out that the primitive behavior of individual entities may remain the same over time. However, if the entities of a complex system are able to adapt, the primitive behavior of entities is bound to be different over time. As a result, different types of complex behavior may be emerged. Moreover, emergent behavior may not arise only through interactions among individual entities. It can also arise through interactions among groups of entities.

Let us take an ant colony as an example to illustrate the above behaviors. Food foraging is an individual task as well as a group task [Goss et al., 1990]. Thus, the *wandering around* of ants is an example of purposeful behavior. Their convergence on a certain food source is an example of emergent behavior. Ants start off with some kind of random walk in the absence of any information about a food source. While wandering, ants lay some quantities of pheromone along their paths. Once a food source is found, more ants will gradually follow the path between the food source and the nest, and consequently more pheromone will be laid along this path. More pheromone will in turn recruit more ants. This process acts as a positive feedback loop, until the food source is exhausted and the pheromone evaporates. This is an example of emergent purposeful behavior.

1.2.2   Autonomy Defined

According to the American Heritage Dictionary of the English Language, autonomy is defined as the condition or quality of being (1) autonomous, independence, (2) self-government or the right of self-government, self-determination, and self-directed. All these conditions or qualities relate to freedom from control by others with respect to primitive behavior. In the field of artificial intelligence, autonomy has been one of the key elements in many research subfields, such as intelligent agents [Jennings and Wooldridge, 1996].

The above is a general definition of autonomy. In what follows, we will define the specific notion of autonomy in the context of AOC, i.e., entity autonomy, synthetic autonomy, emergent autonomy, and computational system autonomy.

Definition 1.5 (Entity autonomy) *Autonomy of an entity refers to its condition or quality of being self-governed, self-determined, and self-directed. It guarantees that the primitive behavior of an entity is free from the explicit control of other entities.*

The above definition is an endogenous view of autonomy. In other words, the primitive behavior of an entity is protected from the influence of others in a way similar to that of an object in the software engineering sense. However, only direct perturbation is prohibited; indirect influence is allowed and encouraged. An underlying assumption is that all entities are able to make decisions for themselves, subject to information availability and self-imposed constraints.

As inspired by the autonomy of entities in natural complex systems, AOC aims at building multi-entity systems where entities are equipped with synthetic autonomy.

Definition 1.6 (Synthetic autonomy) *Synthetic autonomy of an entity is an abstracted equivalent of the autonomy of an entity in a natural complex system. An entity with synthetic autonomy is the fundamental building block of an autonomy oriented computing system.*

A computational system, built from computational entities with synthetic autonomy, exhibits emergent (purposeful) behavior. Correspondingly, we can define emergent autonomy as follows:

Definition 1.7 (Emergent autonomy) *Emergent autonomy is an observable, self-induced condition or quality of an autonomy oriented computing system that is composed of entities with synthetic autonomy.*

A computational system can be described at different levels of abstraction. If a human society is to be modeled as a computational system, abstraction can possibly occur at several levels: population, individual, biological system, cell, molecule, and atom. Note that entity autonomy, synthetic autonomy, and emergent autonomy according to Definitions 1.5-1.7 are present at all these levels. The autonomy obtained at a lower level, say, the cell level, is the foundation of the autonomy at a higher level, say, the biological system level. This multi-level view of autonomy encompasses Brooks' subsumption architecture [Brooks, 1991] in that complex behavior can be built up from multiple levels of simpler, and relatively more primitive, behavior.

Based on the above definitions, autonomy in the context of a computational system can be stated as follows:

Definition 1.8 (Computational system autonomy) *Autonomy in a computational system, built from computational entities with synthetic autonomy, refers to conditions or qualities of having self-governed, self-determined, and self-directed computational entities that exhibit emergent autonomy.*

1.3. General AOC Approaches

AOC contains computational algorithms that employ autonomy as the core of complex systems behavior. They aim at reconstructing, explaining, and predicting the behavior of systems that are hard to be modeled using top-down approaches. Local interaction among autonomous entities is not only the 'glue' that helps entities form a coherent AOC system, but also the primary driving force of AOC. An abstracted version of some natural phenomenon is the starting point of AOC so that the problem at hand can be recast.

Formulating an AOC system involves an appropriate analogy that normally comes from nature. Employing such an analogy requires identification, abstraction, and reproduction of a certain natural phenomenon. The process of abstraction inevitably involves certain simplification of the natural counterpart. For example, the commonly used version of the genetic algorithm [Holland, 1992] in the family of evolutionary algorithms simplifies the process of sexual evolution to selection, recombination, and mutation, without explicit identification of male and female. Evolutionary programming [Fogel et al., 1966] and evolution strategy [Schwefel, 1981] are closer to asexual reproduction with the addition of constraints on mutation and the introduction of mutation operator evolution, respectively. Despite these simplifications and modifications, evolutionary algorithms capture the essence of natural evolution and are proven global optimization techniques.

According to their specific objectives, AOC systems can be developed using one of the three general approaches:

1. **AOC-by-fabrication** aims at replicating and utilizing certain self-organized collective behavior from the real world to form a general purpose problem solver. The working mechanism is more or less known and may be simplified during the modeling process. Research in artificial life is related to this AOC approach up to the behavior replication stage. Nature inspired techniques, such as the genetic algorithm (GA) and the ant colony system, are typical examples of such an approach.

2. **AOC-by-prototyping** attempts to understand the working mechanism underlying a complex system to be modeled. To do so, AOC-by-prototyping characterizes a group of autonomous entities and simulates their observed behavior. A manual trial-and-error process is employed to achieve an artificial system as vivid as possible. Examples of this approach include the study of Internet ecology, traffic jams, Web log analysis, etc..

3. **AOC-by-self-discovery** aims to automatically discover a solution to the problem at hand. The trial-and-error process of the AOC-by-prototyping approach is replaced by an autonomous process in the system. In other words, the difference measure between the desired emergent behavior and the current emergent behavior of the system in question becomes part of the feedback that affects the primitive behavior of an entity. Some evolutionary algorithms that exhibit self-adaptive capabilities are examples of this approach.

1.4.   AOC as a New Computing Paradigm

In the history of computing, there are several computing or programming paradigms worth noting, namely, imperative, functional, logic, and object oriented paradigms. The imperative paradigm embodies computation in terms of a program state and statements that change the program state. An imperative program is a sequence of commands for a computer to perform. The functional paradigm views computation as a process for evaluating a group of mathematical functions. In contrast to the imperative paradigm, it emphasizes the evaluation of functional expressions rather than the execution of commands. Both the imperative and the functional paradigms are to implement a mapping between inputs and outputs. The logic paradigm is to implement a general relation. It describes a set of features that a solution should have rather than a set of steps to obtain such a solution. The object oriented programming (OOP) paradigm is built on the imperative paradigm. It can be viewed as an extension of the imperative paradigm. It encapsulates variables and their operations as classes. As inspired by the object oriented computing paradigm, Shoham proposed an agent oriented programming (AOP) paradigm [Shoham, 1993], where the basic element is an agent characterized by a group of mental parameters, such as beliefs, commitments, and choices.

Unlike previous paradigms, AOC focuses on modeling and developing systems with autonomous entities, in an attempt to solve hard computational problems and to characterize complex systems behavior. In AOC, the basic element is an autonomous entity. The core concept of AOC is the autonomy of entities, which means that entities locally determine their behavior by themselves, and no global control mechanism exists. This is similar to the encapsulation idea of a class in OOP, i.e., the values of its member variables can only be modified by itself.

Table 1.1 briefly compares three paradigms: object oriented programming (OOP), agent oriented programming (AOP), and autonomy oriented computing (AOC). In what follows, we will elaborate their essential differences.

1.4.1 Basic Building Blocks

First of all, it is interesting to note the difference in their basic elements. In OOP, the basic elements are objects embodied by encapsulated variables and corresponding operations. In AOP, the basic elements are agents augmented with mental parameters. In AOC, the basic elements are autonomous entities and their environment. AOC represents and solves a computing problem in a bottom-up fashion. It involves a group of autonomous entities and an environment in which entities reside. An autonomous entity is characterized by its internal states and goals, and provided with an evaluation function, primitive behavior, and behavioral rules. Here, by 'autonomous' we mean that an entity behaves and makes decisions on the changes of its internal states, without control from other entities or a 'commander' outside an AOC system. An entity in AOC does not have mental parameters as in AOP.

Table 1.1. A comparison among object oriented programming (OOP), agent oriented programming (AOP) [Shoham, 1993], and autonomy oriented computing (AOC).

	Object Oriented Programming (OOP)	Agent Oriented Programming (AOP)	Autonomy Oriented Computing (AOC)
Basic element	object	agent	autonomous entity and environment
Characterization of a basic element	member variables and member functions	beliefs, decisions, capabilities, and obligations	states, evaluation function, goals, primitive behavior, and behavioral rules
Interaction	inheritance and message passing among objects	messages among agents, including inform, request, offer, promise, decline, etc.	(1) interaction between entities and their environment and (2) direct or indirect interaction among entities
Computation	message passing and response methods	message passing and response methods	(1) aggregation of behavior and interaction and (2) self-organization in autonomous entities
Suitability	(1) systems modeling and (2) computation based on reusable codes	(1) developing distributed systems and (2) solving distributed problems [Kuhnel, 1997]	(1) solving hard computational problems and (2) characterizing complex systems behavior
Implementation of functionality	member functions	mental state transitions	primitive behaviors

1.4.2   Computational Methodologies

Now let us examine the computational philosophies of different paradigms. In both OOP and AOP, computation is embodied as a process of message passing and response among objects or agents. In AOP, computation involves certain techniques from Artificial Intelligence, such as knowledge representation, inference, and reasoning mechanisms. AOP is suitable for developing distributed systems (e.g., work flow management) and solving distributed problems (e.g., transport scheduling) [Kuhnel, 1997].

In AOC, computation is carried out through the self-organization of autonomous entities. Entities directly or indirectly interact with each other or with their environment in order to achieve their respective goals. As entities simultaneously behave and interact, their outcomes will be nonlinearly aggregated. In an AOC system, the local behavior of entities will be fine-tuned based on negative feedback from the performance of the system. As a result, the non-linear aggregation will be biased to produce certain desired complex systems behaviors or states (i.e., goals). For instance, in the case of problem solving, the emergent system behavior or state corresponds to a solution to a problem at hand.

Generally speaking, AOC works in a bottom-up manner, somewhat like the working mechanism of nature. This is the reason why AOC is well suitable for solving hard computational problems and for characterizing complex systems behavior.

1.5. Related Areas

In the preceding sections, we have presented the key concepts of AOC and have compared it with other computing paradigms. In what follows, we will list some of the existing research areas that are, to a certain extent, related to AOC, and mention their basic distinctions in computational objectives and principles.

- Artificial life (ALife) emphasizes the simulation of life in a computer setting. It, therefore, falls short of its use as a computational approach to problem solving. On the other hand, AOC does not necessarily need to exactly reproduce lifelike behavior, as natural phenomena are usually abstracted and simplified.

- Agent-based simulation (ABS) shares a similar objective with AOC-by-prototyping – finding explanations to observed phenomena. Again, there is no computational problem to be solved in ABS.

- Self-organized criticality (SOC) [Jensen, 1998] was proposed to model certain natural phenomena, such as avalanches, sand pile, rice pile, droplet formation, earthquakes, and evolution. Specifically, SOC "combines [...] [the] concepts [of] self-organization and critical behavior to explain [...] complexity" ([Jensen, 1998], p.2). It studies the critical point where a system will transit from order to chaos, and from stability to avalanche.

- Studies on multi-agent systems for distributed decision making [Durfee, 1999, Sandholm, 1999] attempt to handle computational tasks by delegating responsibilities to groups of agents. These agents are usually heterogeneous entities, and different groups have different roles. For example, in the Zeus collaborative agent building framework, agents are divided into utility agents, such as name server, facilitator, and visualizer, and domain level agents [Nwana et al., 1998]. The behavior of an individual agent is preprogrammed and can sometimes be complex. Complicated issues, such as negotiation and coordination, are of paramount importance. All these are usually part of the systems design, and hence require human interventions.

- Another example of multi-entity computation is *ecology of computation* [Hogg and Huberman, 1993, Huberman, 1988], where a population of heterogeneous computational systems share partial solutions to a common problem at hand. These individual problem solvers tackle the problem with different methods and, therefore, have different internal representations of the problem and solution. While it has been shown that this approach is efficient in solving a problem, the coordination of problem solvers needs to be carefully articulated so that different internal representations can be properly translated.

- Distributed constraint satisfaction problem (distributed CSP) [Yokoo et al., 2001, Yokoo et al., 1998, Yokoo and Hirayama, 2000] was proposed to solve CSPs in a distributed manner. Specifically, the asynchronous backtracking algorithm assigns a variable to an agent. A directed graph, called constraint network, is constructed to represent the relationships between variables. The asynchronous weak-commitment search algorithm enhanced the above algorithm by adding a dynamical priority hierarchy. In the event of conflicts, the hierarchy structure is changed so that a suboptimal solution can be found first before incrementally arriving at a final solution.

  In distributed CSP, agents employ direct communications to coordinate the assignments to their respective variables. Generally speaking, direct communication in a large-scale multi-agent system is time-consuming. Hence, in an AOC system, autonomous entities utilize indirect communication through their environment. Furthermore, AOC is inspired by complex systems where numerous entities self-organize themselves through nonlinear interactions and aggregations, and gradually emerge certain complex behaviors. Since it is based on the idea of self-organization, AOC is suitable for large-scale, highly distributed problems.

- Swarm intelligence [Bonabeau et al., 1999, Bonabeau et al., 2000] is a demonstration of AOC-by-fabrication. It utilizes a social insect metaphor in problem solving. Unlike AOC, this study does not address the issues of discovering problem solvers or explaining complex systems behavior.

1.6.   Summary

In this chapter, we have defined several fundamental notions underlying the AOC paradigm. We started with some observations of complex systems in natural and man-made systems, such as traffic, stock markets, and sports. We then identified several important characteristics in the basic entities of such complex systems, namely, autonomy, emergence, adaptation, and self-organization. These properties are the hallmarks of AOC as a new bottom-up paradigm for computing. As compared to other existing computing paradigms or research areas, the goals of AOC are unique: *To computationally synthesize emergent purposeful behavior for problem solving and complex systems modeling.*

Also in this chapter, we have introduced three general AOC approaches, i.e., AOC-by-fabrication, AOC-by-prototyping, and AOC-by-self-discovery. AOC-by-fabrication refers to the reproduction of emergent behavior in computation. With some knowledge of an underlying mechanism, an analogy of the mechanism observed from the emergent behavior is used as a general purpose problem solving technique. Synthesizing emergent behavior is not the end, but rather the means, of an AOC algorithm. AOC-by-prototyping is used to understand the emergent behavior and working mechanism of a real-world complex system by hypothesizing and repeated experimentation. The end product of these simulations is a better understanding of, or explanations to, the real working mechanism of the modeled system. AOC-by-self-discovery refers to the AOC algorithms or systems that automatically emerge problem solvers or systems models in the absence of human intervention. In other words, self-adaptive algorithms or systems are desired.

Exercises

1.1 Observe and briefly describe some forms of autonomy in natural or man-made systems. Think about how such forms may have been created in the first place, who are the key players, and what are the determining factors.

1.2 Give examples to highlight the differences between AOC and the process models for the simulation of discrete event systems.

1.3 Summarize the various concepts of behavior and autonomy as defined in this chapter, and then try to establish some inter-relationships among them.

1.4 AOC emphasizes the principle of self-organization. How is this reflected in the three general approaches of AOC?

1.5 With the help of a computing encyclopedia or the Web, write down your own definition of a top-down problem solving approach. What are the major steps in such an approach?

1.6 Using AOC as an example of bottom-up approaches, illustrate their key characteristics. What are the main differences and relationships between bottom-up and top-down systems?

1.7 What are the advantages and disadvantages of top-down and bottom-up problem solving approaches over each other? Show some examples of modeling or computing problems in which top-down approaches are not applicable. When are they applicable?

1.8 Provide an in-depth survey of related references in the fields of computer science and systems science on some of the following concepts that are closely related to certain specific formulations and stages of an AOC approach: autonomy, emergence, adaptation, and self-organization.

1.9 Compare the scope of AOC with those of adaptive computation and complex adaptive systems studies as may be found in the existing literature. Explain how the AOC paradigm unifies such methods.

1.10 By studying the literature on swarm intelligence, explain how swarm intelligence methods can be generalized into an AOC approach.

1.11 Among the computing and programming paradigms mentioned in this chapter, find one application problem for each of them. You should carefully consider the nature of the problem such that it can best demonstrate the strengths of a paradigm adopted.

Chapter 2

AOC at a Glance

2.1. Introduction

AOC approaches share a basic form with many possible variants. As mentioned in the preceding chapter, the basic form draws on the core notion of autonomy, with such characteristics as multi-entity formulation, local interaction, nonlinear aggregation, and self-organized computation. In order to get a better idea on how these characteristics are reflected in handling some familiar computational or engineering problems, let us now take a look at three illustrative examples: constraint satisfaction, image feature extraction, and robot spatial learning (or world modeling). In our illustrations, we will outline the basic autonomy models implemented and highlight the AOC systems performance.

2.2. Autonomy Oriented Problem Solving

Our first illustrated example deals with distributed problem solving. Liu et al. have developed an AOC-based method for solving CSPs [Liu and Han, 2001, Liu et al., 2002]. This method is intended to provide an alternative, multi-entity formulation that can be used to handle general CSPs and to find approximate solutions without too much computational cost.

2.2.1 Autonomy Oriented Modeling

In the AOC-based method, distributed entities represent variables and a two-dimensional lattice-like environment in which entities inhabit corresponds to the domains of variables. Thus, the positions of entities in such an environment constitute a possible solution to the corresponding CSP. The distributed multi-entity system self-organizes itself, as each entity follows its behavioral rules, and gradually evolves towards a global solution state.

Based on two general principles of 'survival of the fittest' – poor performers will be washed out, and 'law of the jungle' – weak performers will be eliminated by stronger ones, the AOC-by-fabrication approach is applied to solve a benchmark constraint satisfaction problem [Han et al., 1999, Liu and Han, 2001].

2.2.2   N-Queen Problem

The n-queen problem aims to allocate n queens on an $n \times n$ chessboard such that no two queens are placed within the same row, column, and diagonal. Based on the constraints of the problem, a model is formulated in the following manner. Each queen is modeled as an autonomous entity in an AOC system and multiple queens are assigned to each row on the chessboard. This is to allow competition among the queens in the same row such that the queen with the best strategy survives. The system calculates the number of violated constraints (i.e., violations) for each position on the chessboard. This represents the environmental information to all queens in making movement decisions, which are restricted to positions in the same row. Queens are allowed for three types of movement. A 'randomized-move' allows a queen to randomly select a new position. A 'least-move' selects a position with the least number of violations. A 'coop-move' promotes cooperation between queens by excluding positions that will attack those queens with which one wants to cooperate. These types of movement are selected probabilistically.

An initial energy is given to each queen. A queen will 'die' if its energy falls below a predefined threshold. Energy will change in two ways. When a queen moves to a new position that violates the set constraint with m queens, it loses m units of energy. This will also cause those queens that attack this new position to lose one unit of energy. The intention is to encourage a queen to find a position with the least number of violations. The 'law of the jungle' principle is implemented by having two or more queens occupying the same position to compete for the occupancy. The queen with the highest energy will win and eliminate the loser(s) by absorbing all the energy of the loser(s).

The above model is able to efficiently solve n-queen problems with up to 7,000 queens using a moderate hardware configuration. Experimental results show that the 'survival of the fittest' principle helps find an optimal solution much more quickly due to the introduction of competition. The randomized-move is indispensable as it helps an AOC system come out of local minima, although giving a high chance of making a randomized-move will lead to chaotic behavior. The probabilities of selecting 'least-move' and 'coop-move' should be comparable and increased with the size of a problem.

2.3.   Autonomy Oriented Search

In our next example, let us consider the following search problem: An environment contains a homogeneous region with the same physical feature. This region is referred to as a goal region. The feature of the goal can be evaluated based on some measurements. Here, the term 'measurement' is taken as a generic notion. The specific quantity that it refers to depends on the nature of applications. For instance, it may refer to the grey level intensity of an image, in the case of image processing. The task of autonomous entities is to search the feature locations of the goal region. Entities can recognize and distinguish feature locations, if encountered, and then decide and execute their reactive behavior.

2.3.1   Autonomy Oriented Modeling

In the AOC-based method, an entity checks its neighboring environment, i.e., small circles as in Figure 2.1(a), and selects its behavior according to the concentration of elements in the neighboring region. If the concentration is within a certain range, the current location satisfies a triggering condition. This activates the reproduction mechanism of the entity.

Taking a border tracing entity for example (see Figure 2.1(a)), if an entity of the border sensitive class reaches a border position, this entity will inhabit at the border and proceed to reproduce both within its immediate neighboring region and inside a large region, as illustrated in Figures 2.1(b) and (c).

2.3.2   Image Segmentation Problem

Image segmentation requires to identify homogeneous regions within an image. However, homogeneity can be at varying degrees at different parts of the image. This presents problems to conventional methods, such as split-and-merge that segments an image by iteratively partitioning heterogeneous regions and simultaneously merging homogeneous ones [Pavlidis, 1992, Pitas, 1993]. An autonomy oriented method has been developed to tackle the same task [Liu et al., 1997]. Autonomous entities are deployed to the two-dimensional representation of an image, which is considered as the search space of entities. Each entity is equipped with an ability to assess the homogeneity of a region within a predefined locality. Specifically, homogeneity is defined by the relative contrast, regional mean, and region standard deviation of the grey level intensity. When an autonomous entity locates a homogeneous region within the range of the pixel at which it presently resides, it breeds a certain number of offspring entities and delivers them to its local region in different directions. On the other hand, when a heterogeneous region is found, an entity will diffuse to another pixel in a certain direction within its local region.

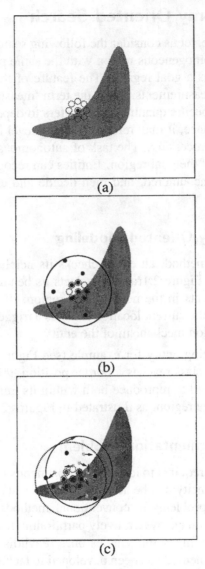

Figure 2.1. An illustration of the behavior of autonomous entities. (a) As an entity, which is marked as a solid circle, moves to a new location, it senses its neighboring locations, marked by dotted circles in this example. Specifically, it counts the number of locations at which the grey level intensity is close to that of the entity's current location. (b) When the count reaches a certain value, it is said that a triggering condition has been satisfied. This is in fact the case in our illustrative example, as the location of the entity is right next to the border of a shaded region. Thus, the entity will asexually self-reproduce some offspring entities within its local region. (c) At the following steps, the offspring will diffuse to new locations. By doing so, some of them will encounter new border feature locations as well and thereafter self-reproduce more entities. On the other hand, the entities that cannot find any border feature locations after a given number of diffusion steps will be automatically turned off [Liu, 2001].

Through breeding behavior, an entity distributes its newly created offspring into the region that is found to be homogeneous, so that the offspring entities are more likely to find extensions to the current homogeneous region. Apart from breeding, an entity will also label an pixel that is found to be in a homogeneous region. If an autonomous entity fails to find a homogeneous region during its lifespan (a predefined number of steps) or wanders off the search space during diffusion, it will be marked as an inactive entity.

In summary, the stimulus from pixels will direct autonomous entities to two different behavioral tracts: breeding and pixel labeling, or diffusion and decay. The directions of breeding and diffusion are determined according to their respective behavioral vectors, which contain weights (between 0 and 1) of all possible directions. The weights are updated by considering the number of successful siblings in the respective directions. An entity is considered to be successful if it has found one or more pixels that are within a homogeneous region. This method of direction selection is somewhat similar to herding behavior that only considers local information. A similar technique has been applied to feature extraction tasks, such as border tracing and edge detection [Liu and Tang, 1999]. A more difficult task where an image contains different homogeneous regions has been successfully handled by deploying autonomous entities with different homogeneity criteria.

2.3.3 An Illustrative Example

In order to examine the above autonomy oriented method in the simultaneous detection of significant image segments, Liu et al. [Liu, 2001, Liu et al., 1997, Liu and Tang, 1999] have conducted several experiments in which various classes of entities are defined and employed to extract different homogeneous regions from an image, such as the example given in Figure 2.2 ($t = 0$). For this image segmentation task, 1,500 entities, evenly divided into three classes, are randomly distributed over the given image. Figure 2.2 presents a series of intermediate steps during the collective image segmentation. Figure 2.2 ($t = 50$) gives the resultant markers as produced by the different classes of entities.

2.3.4 Computational Steps

In the AOC-based image segmentation, the computational steps required can be estimated by counting how many active entities are being used over time (i.e., the entities whose ages do not exceed a given life span). For the above mentioned collective image segmentation task, we have calculated the number of active entities in each class that have been involved over a period of 50 steps, as given in Table 2.1. It can readily be noted that the total number of

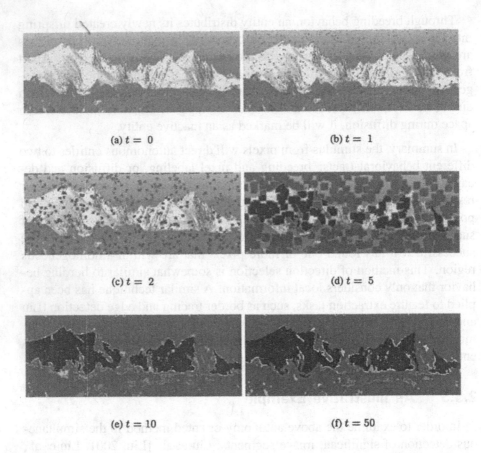

(a) $t = 0$ (b) $t = 1$

(c) $t = 2$ (d) $t = 5$

(e) $t = 10$ (f) $t = 50$

Figure 2.2. Segmenting a landscape image that contains three complex-shaped homogeneous regions [Liu, 2001].

active entities (i.e., computational steps) involved in extracting a homogeneous region is less than the size of the given image, $526 \times 197 = 103,622$.

Table 2.1. The number of active entities involved in extracting the homogeneous regions of a landscape image.

Class	# of active entities used (time step $= 1 \sim 50$)
Class-1	47,037
Class-2	75,473
Class-3	48,837

2.4. Autonomy Oriented Learning

Our final example demonstrates how AOC can be embodied in a group of distributed autonomous robots to perform a task of collective spatial learning (or world modeling).

Ant colonies are able to collect objects, such as food or dead ants, and place them in particular places. Collective behavior in a complex system offers the possibilities of enhanced task performance, increased task reliability, and decreased computational cost over traditional complex systems. Much work to date in collective robotics focuses on limited cases, such as flocking and foraging. Typical entities in those studies either use manually built (non-learning) controllers [Balch and Arkin, 1995], or perform a learning task in simulated [Balch, 1997] or relatively simple physical environments [Mataric, 1994]. One way to generate robust collective behavior is to apply biologically inspired adaptive algorithms at a team level. In such a case, the environment plays a central role in triggering a certain basic behavior at any given time. It draws on the idea of providing robots with a range of primitive behaviors and letting the environment determine which behavior is more suitable as a response to a certain stimulus. The integration of learning methods can significantly contribute to the performance of a team of self-programming robots for some predefined tasks. These individual robots can automatically program their task-handling behavior to adapt to dynamical changes in their task environment in a collective manner.

2.4.1 World Modeling

Liu and Wu have developed an AOC-based method for collective world modeling with a group of mobile robots in an unknown, less structured environment [Liu and Wu, 2001]. The goal is to enable mobile robots to cooperatively perform a map building task with fewer sensory measurement steps, that is, to construct a potential field map as efficiently as possible. The following issues are addressed in developing the proposed world modeling method:

- How to formally define and represent the reactive behavior of mobile robots and the underlying adaptation mechanisms to enable the dynamical acquisition of collective behavior?

- How to solve the problem of collective world modeling (i.e., potential field map building) in an unknown robot environment based on self-organization principles?

The artificial potential field (APF) theory states that for any goal directed robot in an environment that contains stationary or dynamically moving obstacles, an APF can be formulated and computed by taking into account an attractive pole at the goal position of the robot and repulsive surfaces of the obstacles. Using APF, any dynamical changes in the environment can be modeled by updating the original artificial potential field. With APF, a robot can reach a stable configuration in its environment by following the negative gradient of its potential field.

An important challenge in the practical applications of the APF methodology is that evolving a stable APF is a time consuming learning process, which requires a large amount of input data coming from the robot-environment interaction. The distributed self-organization method for collective APF modeling with a group of mobile robots begins with the modeling of local interactions between the robots and their environment, and then applies a global optimization method for selecting the reactive motion behavior of individual robots in an attempt to maximize the overall effectiveness of collectively accomplishing a task.

The main idea behind self-organization based collective task handling is that multiple robots are equipped with a repository of behavioral responses in such a way as to create some desirable global order, e.g., the fulfillment of a given task. For instance, mobile robots may independently interact with their local environment. Based on their performance (e.g., distributed proximity sensory measurements), some global world models of an unknown environment (i.e., global order) can be dynamically and incrementally self-organized.

2.4.2 Self-Organization

In the case of collective world modeling, self-organization is carried out as follows: Suppose that a robot moves to position p_0 and measures its distances to the surrounding obstacles of its environment in several directions (n). These measurements are recorded in a sensing vector, $\mathcal{S}_0 = [d_1^0, d_2^0, \cdots, d_i^0, \cdots, d_n^0]$, with respect to position p_0 where d_i^0 denotes the distance between position p_0 and an obstacle sensed in the ith direction. The robot will then associate this information to its adjacent positions in the environment by estimating the proximity values in the neighboring positions. The estimated proximity of any position p_j inside the neighboring region of p_0 to a sensed obstacle will be calculated as follows: $\hat{d}_i^j = d_i^0 - \rho_j \cdot cos\beta$ $(i = 1, 2, \cdots, n)$, where $\beta = \alpha_0^{(i)} - \alpha_j$. $\alpha_0^{(i)}$ and α_j denote the polar angle of the sensing direction and that of position p_j, respectively. $\hat{d}_i^j$ is an estimate for p_j based on the ith direction sensing value. d_i^0 is the current measurement taken from p_0 in the ith direction. Thus, the estimated proximity values for position p_j can be written as: $\hat{\mathcal{S}}_j = \left[\hat{d}_1^j, \hat{d}_2^j, \cdots, \hat{d}_i^j, \cdots, \hat{d}_n^j\right]$. Figure 2.3 illustrates the distance association scheme.

Next, we define a confidence weight for each element of $\hat{\mathcal{S}}_j$, that is, a function of the distance between a robot and position p_j, or specifically, $w_j =$

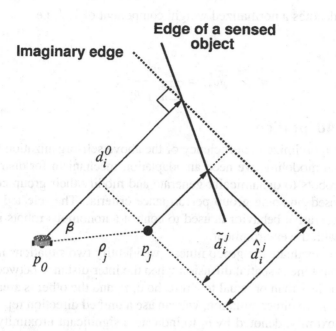

Figure 2.3. An illustration of the distance association scheme [Liu, 2001].

$e^{-\eta \rho_j^2}$, where η is a positive constant; ρ_j is the distance between the robot and position p_j.

The potential field estimate at position p_j is then computed as follows:

$$\hat{U}_j^t = \sum_{i=1}^{n} e^{-\lambda \hat{d}_i^j}, \qquad (2.1)$$

where λ is a positive constant. Thus, at time t, a set of potential field estimates, $\Omega_t^j = \{\hat{U}_j^{t_1}, \hat{U}_j^{t_2}, \cdots, \hat{U}_j^{t_i}, \cdots, \hat{U}_j^{t_k}\}$, can be derived by k robots with respect to position p_j, that is,

$$\Omega_t^j \leftarrow \Omega_{t-1}^j \bigcup \mathcal{Q}, \qquad (2.2)$$

where Ω_{t-1}^j denotes a set of potential field estimates for position p_j at time $t-1$, and $\mathcal{Q} = \hat{U}_j^{t_k}$, where subscript k indicates that the potential value is estimated based on the measurement of the kth robot. Ω_t^j is associated with a confidence weight set: $W_t^j = \{w_j^{t_1}, w_j^{t_2}, \cdots, w_j^{t_i}, \cdots, w_j^{t_k}\}$.

Hence at time t, an acceptable potential field value can readily be calculated as follows:

$$U_j^t = \begin{cases} \hat{U}_j^{t_i}, & \exists i \in [1, k], w_j^{t_i} = 1, \\ \sum_{i=1}^{k} \hat{U}_j^{t_i} \cdot \bar{w}_j^{t_i}, & \text{otherwise,} \end{cases} \qquad (2.3)$$

where $\bar{w}_j^{t_i}$ denotes a normalized weight component of W_t^j, i.e.,

$$\bar{w}_j^{t_i} = \frac{w_j^{t_i}}{\sum_{n=1}^{k} w_j^{t_n}}. \tag{2.4}$$

2.4.3   Adaptation

In order to optimize the efficiency of the above self-organization based collective world modeling, we need an adaptation mechanism for distributed autonomous robots to dynamically generate and modify their group cooperative behavior based on some group performance criteria. The selected (i.e., high fitness) cooperative behavior is used to control autonomous robots in their interactions with the environment.

In order to evaluate the group fitness, we identify two situations involved in the evolution: One is spatial diffusion when the inter-distance between robots i and j, σ_{ij}, is less than or equal to a threshold, η, and the other is area coverage when $\sigma_{ij} > \eta$. In either situation, we can use a unified direction representation of robot proximity, denoted by θ_i to indicate a significant proximity direction of all proximity stimuli to robot i. Having identified these two situations in group robots, we can reduce the problem of behavior evolution into that of acquiring two types of individual reactive motion behavior: One for spatial diffusion and the other for area coverage, respectively. Both types of reactive behavior respond to proximity stimuli as defined in terms of a unified significant proximity direction.

The fitness function will consist of two terms: One is called general fitness, denoted by f_g, and the other is called special fitness, denoted by f_s. The general fitness term encourages group robots to explore the potential field in new, less confident regions, and at the same time, avoid repeating the work of other robots. It is defined as follows:

$$f_g = \prod_{i=1}^{m} \left\{ (1 - \max\{w_i^{t_k}\}) \prod_{j=1}^{m_e} \sqrt[4]{\sigma_{ij} - \delta} \right\}, \tag{2.5}$$

where $\max\{w_i^{t_k}\}$ denotes the maximal confidence weight corresponding to the position of robot i. m denotes the number of robots that are grouped together during one evolutionary movement step (of several generations). m_e denotes the number of robots that do not belong to m and have just selected and executed their next behavior. σ_{ij} denotes the distance between robots i and j, which is greater than a predefined distance threshold, δ.

Two special fitness terms will be defined corresponding to the performance of spatial diffusion and area coverage:

$$\text{spatial\_diffusion: } f_{s1} = \prod_{i=1}^{m_d-1} \prod_{j=i+1}^{m_d} \sqrt{\sigma_{ij} - \eta}, \qquad (2.6)$$

and

$$\text{area\_coverage: } f_{s2} = \frac{\sqrt{\Delta \mathcal{V}}}{\prod_{i=1}^{m_c} \zeta_i}, \qquad (2.7)$$

where m_d denotes the number of spatially diffusing robots whose inter-distances σ_{ij} have become greater than the distance threshold, η. $\Delta \mathcal{V}$ denotes the total number of positions visited by a group of m_c area-covering robots based on their selected motion directions. ζ_i denotes a significant proximity distance between robot i and other robots in the environment.

2.5.   Summary

So far, we have provided three illustrative examples: constraint problem solving, distributed search, and spatial learning. We have stated the basic problem requirements and showed the ideas behind the AOC solutions, ranging from their formulations to the emergence of collective solutions through self-organization.

From the illustrations, we can note that using an AOC-based method to solve a problem is essentially to build an autonomous system, which usually involves a group of autonomous entities residing in an environment. Entities are equipped with some simple behaviors, such as move, diffuse, breed, and decay, and one or more goals (e.g., to locate a pixel in a homogeneous region). In order to achieve their goals, entities either directly interact with each other or indirectly interact via their environment. Through interactions, entities accumulate their behavioral outcomes and some collective behaviors or patterns emerge. Ideally, these collective behaviors or patterns are what we are expecting, i.e., solutions to our problems at hand.

Exercises

2.1 Provide a conceptual blueprint for a potential AOC programming language that can support the applications as mentioned in this chapter. What will be its constructs? What will be the key characteristics and requirements of its operations?

2.2 In order to evaluate computing languages or environments for AOC, what will be your suggested criteria? What will be your suggested benchmark problems?

2.3 AOC offers a new way of tackling complexity, whether in problem solving or in complex systems modeling, by utilizing localized, autonomous and yet low-cost (computationally and physically speaking), and self-organized entities. Based on the illustrative examples given in this chapter, try to suggest and develop some other alternative models of AOC, as inspired by nature, for solving the same or different problems.

2.4 Identify from the above solved problems the real benefits of taking this route to complexity.

2.5 The example on world modeling in this chapter utilizes self-organized collective behavior in a multi-robot system to model an unknown environment. Think and propose a similar solution to the problem of multi-robot navigation.

2.6 Can you summarize the similarity in computational ideas between the image segmentation example and the world modeling example? (for instance, both have been treated as distributed problem solving).

2.7 The chapter illustrates a search (optimization based) strategy for image segmentation and feature detection. From a computer vision point of view, compare this strategy with some traditional search based segmentation algorithms, and evaluate their performances with other computer vision segmentation benchmarks.

Chapter 3

Design and Engineering Issues

3.1.   Introduction

In the preceding chapters, we have provided the basic concepts and illustrations of AOC in problem solving. From a design and engineering point of view, we still face several unsolved issues, namely, what is an autonomous entity in an AOC system generally composed of? What are the functional modules of an entity? What are the inter-relationships among autonomous entities? What steps are involved in engineering an AOC system? What features should be demonstrated from such engineered AOC systems? In what follows, we will address these design and engineering issues.

3.2.   Functional Modules in an Autonomous Entity

Generally speaking, an AOC system contains a group of autonomous entities. Each autonomous entity is composed of a detector (or an array of them), an effector (again, there can be an array of them), and a repository of local behavioral rules. Figure 3.1 presents a schematic diagram illustrating the functional modules in an autonomous entity.

As shown in Figure 3.1, the role of a detector in an entity is to receive information from its neighbors as well as its local environment. For instance, in the AOC-based image segmentation, this information is the grey-scale intensities of the neighboring positions of an entity. In the case of a bird flocking simulation, this information includes the speeds of birds, directions in which they are heading, and the distances between the birds in question. Details of the content and format of this information need to be defined according to the system to be modeled or the problem to be solved.

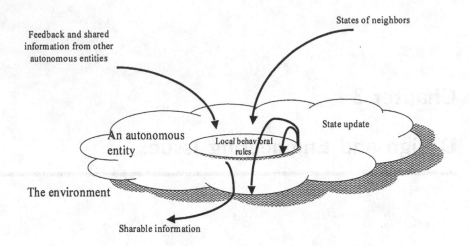

Figure 3.1.   Functional modules in an autonomous entity.

Here, the notion of neighbor can be defined in terms of position (e.g., the bird in front, to the left, and to the right), distance (e.g., a radial distance of two grids), or both (i.e., the birds up to 2 grids in front). Environmental information conveys the status of a certain feature, for example, the presence of food, which is of interest to an autonomous entity. The environment can help carry sharable information.

The effector of an autonomous entity refers to the device for expressing behaviors. These behaviors can make a change either in the internal state of an entity or in the environment where the entity inhabits. An important role of the effector, as a part of the primitive behavior model, is to facilitate implicit information sharing between autonomous entities.

Central to an autonomous entity are the behavioral rules that govern how the entity should act or react to the information collected by its detector from the environment and its neighbors. These rules decide what state this entity should change to and what information this entity should release via its effector to the environment. An example of sharable information is the pheromone in an ant system. This information is untargeted and the communications among ants via the environment are undirected; any ant can pick up the information and react according to its own behavioral rules.

In order to adapt themselves to a dynamically changing environment without being explicitly told in advance, autonomous entities need to modify their behavioral rules over time. This is the learning capability of autonomous entities.

It is worth noting that randomness plays a part in the decision making process of an autonomous entity, despite the presence of a behavioral rule set.

This is to allow an autonomous entity to explore uncharted territories even in the presence of mounting evidence that it should exploit a certain path. On the other hand, randomness helps an autonomous entity resolve conflicts in the presence of decision making uncertainty, and avoid getting stuck in local optima.

As one of the main components in an AOC system, an environment usually plays three roles. First, it serves as the domain in which autonomous entities roam. This is a static view of the environment. Secondly, the environment can also act as a 'notice board' where autonomous entities can post and read their sharable information. In this sense, the environment can also be regarded as an indirect communication medium among entities. This is a dynamical view of the environment. For example, StarLogo [Resnick, 1994] has patches in the environment which knows how to grow food or evaporate ants' pheromone. Sometimes the environment provides feedback to autonomous entities regarding their behavior. For example, in the n-queen constraint satisfaction problem, the environment can tell a queen how many constraints are violated in its neighborhood after it takes a move. This, in effect, translates the global goal of the whole AOC system to a local goal of individual entities. Thirdly, the environment keeps a central clock that helps synchronize the behaviors of all autonomous entities, if necessary.

3.3.   Major Phases in Developing AOC Systems

In the second part of this book, we will present a number of case studies on AOC applications. These applications cover not only constraint satisfaction and optimization problem solving, but also complex systems behavior characterization. AOC systems in these case studies share several commonalities.

First, there are a collection of autonomous entities. If AOC is used to solve a particular problem, the right level of abstraction has to be chosen so that autonomous entities can be identified. This does not exclude the repeated application of this technique to suit the need of the specific problem that can be best modeled by multiple levels of abstraction. Secondly, there are some relationships between autonomous entities in the form of constraints, such as limitations on the position inhabitable by a queen in an n-queen problem. Thirdly, a performance measurement is available to assess the quality of any solution. It can be used in AOC as a guideline to direct the behavior of autonomous entities.

AOC can be viewed as a methodology for engineering a computing system to solve hard computational problems or model complex systems. In general, developing an AOC system or formulating an AOC algorithm often involves three major phases (see Figure 3.2). The first phase, natural system identification, can be viewed as the precursor to actual system modeling. It concerns the selection of an appropriate analogy in the natural or physical world. It involves

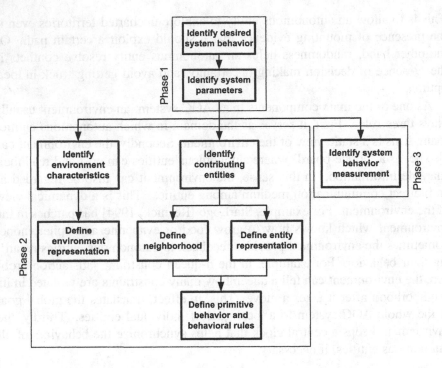

Figure 3.2. The major phases in developing an AOC system.

two specific tasks: 'identify desired behavior' and 'identify system parameters'. Choosing a right analogy is key to the success of an AOC system and the right system usually presents itself through its behavior. Once an appropriate analogy is chosen, details, such as the number of entities to run and the length of time to run a simulation, need to be decided.

The second phase, artificial system construction, involves all elements in an AOC system. This phase is divided into two major subphases: autonomous entity modeling and environment modeling. The 'identify contributing entities' task is the first and the most important task in this phase. Designers are required to choose the level of detail to be modeled, which is appropriate to the problem at hand. The 'define neighborhood' task defines a certain measurement (e.g., positions and distance) in the environment. Within the neighborhood of an entity, local interactions can occur and local information can be collected. The 'define entity representation' task handles how to characterize an autonomous entity, including its internal states and goals etc. The last task concerning entities, 'define primitive behavior and behavioral rules', defines the ways in which an autonomous entity reacts to various information it has collected within its neighborhood and the ways in which it adapts its primitive behavior and behavioral rules.

The tasks that concern the environment are 'identify environment characteristics' and 'define environment representation'. The former task concerns the role that the environment plays in conveying the information shared between autonomous entities. The latter task addresses the characterization of the environment.

The third phase, performance measurement, concerns the evaluation criteria for comparing the artificial system manifested by an AOC system with its natural counterpart. The measurement provides an indication to modify the current set of individual behaviors and behavioral rules. The end of this phase triggers the next cycle of the AOC modeling, if necessary, and involves modifications to some or all AOC elements defined in the previous cycle.

3.4. Engineering Issues

The above mentioned three phases can be implemented with varying degrees of human involvement, from detailed engineering work to an automated process, leading to AOC systems capable of achieving various problem solving and systems modeling objectives.

From an engineering point of view, AOC systems differ in at least five aspects:

1. **Knowledge of working mechanism:** It is often the case in AOC that analogies are drawn from the natural or physical world. Therefore, the working mechanism becomes the key basis of modeling.

2. **Designer's involvement:** Having detailed knowledge of the working mechanism means more detailed work by a designer in implementing an AOC system or algorithm.

3. **Uncertainty in results:** Uncertainty in the outcome of an AOC system varies according to the knowledge of the actual working mechanism.

4. **Computational cost:** Computational cost refers to the time and space complexity of an algorithm. Generally speaking, the computational cost of a general AOC algorithm is higher than that of a customized AOC algorithm for a particular problem.

5. **Time in process modeling:** We refer to the time spending on building an AOC system as the time in process modeling. It hitches on the complexity of the problem to be solved or the system to be modeled as well as the degree of understanding of the problem or system.

Figure 3.3 shows the relative rank of the general AOC approaches according to the above five criteria.

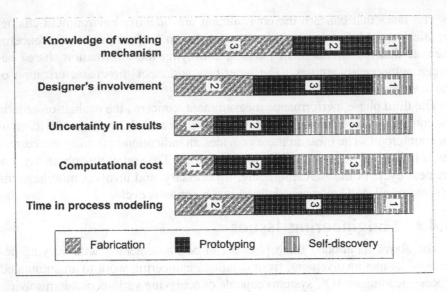

Figure 3.3.   Ranking AOC approaches according to their engineering requirements: 1 being the easiest and 3 the hardest.

Following the AOC-by-fabrication approach requires detailed knowledge of the working mechanism and a high degree of a designer's involvement in implementing a corresponding AOC system. On the other hand, an AOC system built in this way has the lowest computational cost and it is safe to say that it will work. The overall time needed to build such an AOC system is relatively long.

Building an AOC-by-self-discovery system, on the other hand, does not heavily rely on detailed knowledge about the working mechanism because it is usually unknown. A designer does not spend too much time on crafting the system. As a result, an AOC-by-self-discovery system requires higher computational cost than an AOC-by-fabrication system. It naturally takes a long time to achieve its task. It follows that the risk of failure (i.e., not able to find a solution to the problem to be solved or an appropriate model of the system to be modeled) is the highest with an AOC-by-self-discovery system.

The level of difficulty in constructing an AOC-by-prototyping system is between the above two, as some information is already known. As an trial-and-error process is involved in the AOC-by-prototyping approach, an AOC-by-prototyping system requires the most involvement of a designer. Accordingly, more time is generally needed to build such a system. In addition, an AOC-by-prototyping system incurs a higher risk of failing to find a solution.

3.5. Features and Characteristics of AOC Systems

AOC systems that are designed and engineered following the outlines provided in the preceding sections will exhibit some most important and interesting features, which can be characterized as follows:

Homogeneity Complex systems are made up of similar individuals. They may differ only in the parameters for characterizing their goals, behaviors, and behavioral rules, but not in their structures. For example, in the AOC-based image segmentation, although entities are classified into three different classes, they are homogeneous in that they differ only in the parameters for describing their goals.

Simplicity The behavior model of each autonomous entity is simple.

Locality The interactions among autonomous entities are strictly local although the notion of locality can be physical as well as logical.

Implicity Another form of interaction comes from the implicit knowledge sharing among autonomous entities via their common environment.

Uncertainty Behavior is not purely deterministic. There is always a certain degree of randomness and uncertainty in the decision making process of entities.

Amplification Desirable behavior is amplified while undesirable behavior is eliminated via mechanisms, such as birth and death. This is also the result of positive feedback from the environment.

Recursion Complex behavior is aggregated from primitive autonomy of entities through iterations. The result of aggregation is emergent autonomy. A system exhibiting such emergent autonomy can serve as the basic element of a more complex system that may show its own emergent autonomy.

Scalability Cutting across all the AOC approaches is the issue of scalability. AOC seeks to take a bold view of scalability requirements, as it envisions an environment able to scale up or down according to the ever changing needs of a dynamical complex system.

Openness New types of autonomous entity can be accommodated seamlessly into an AOC system.

3.6.   Performance Considerations

Any computational algorithms can be assessed with respect to their generality, robustness, completeness, efficiency, and computational cost. Generality measures the applicability of an algorithm to different problem domains. Robustness concerns the sensitivity of an algorithm in terms of its parameter settings. Completeness of an algorithm assesses its ability to search the whole solution space. Efficiency measures the effectiveness of an algorithm to find an optimal solution and how quickly such a solution is found. It is usually reflected in the computational cost of an algorithm. Here, computational cost refers to the requirements on computational cycles and memory space in the process of finding a solution.

AOC is a general problem solving methodology. It is not designed for a particular problem. Therefore, it should highly score on generality. AOC algorithms score equally high on completeness as they are able to effectively cover the 'more promising' areas of the solution space. On the front of efficiency, AOC algorithms can always find a 'good enough' solution within a short period of time. Given enough time, the globally optimal solution can be found. Directly related to this is the computational cost. AOC algorithms may not highly score in this aspect as they usually have a group of elements and require some computational cycles for evaluating each candidate and accumulating enough positive feedback to make a solution stand out. Robustness is usually high but sometimes is hampered by the formulation of a solution as it directly affects the quality measure of a solution and indirectly the progress.

Besides the above considerations, there exist a number of factors that affect the performance of an AOC system. In what follows, we will highlight some of the most critical ones.

Randomness is an important factor in any self-organizing system, such as those formulated according to the AOC principles. On one hand, it helps autonomous entities in an AOC system explore new areas in the corresponding solution space. On the other hand, it introduces a degree of uncertainty regarding the outcome of a simulation. Therefore, it is important to have a concrete assessment of the progress in an AOC system.

Emergence is a property of AOC that is not pre-programmable. Therefore, it is not possible to measure it directly from the parameters that characterize an AOC system and its elements. Wright et al. have suggested a measure of emergence by likening a self-organized system to a set of nonlinear springs and dampers to represent local interactions among autonomous entities [Wright et al., 2000]. They argued that factors, such as spatial and velocity coherence in a flock of birds, are observable behaviors due to the same underlying mechanism. They further argued that there is a strong correlation between the abrupt changes in observable behavior and emergent behavior while system parameters are being changed smoothly. Therefore, these factors need to be considered together, not separately. A Hamilitonian model using Shannon entropy is formulated and a dimensionality measure is defined based on the entropy, which is linked directly with a phase transition hypothesis and has been used as feedback to a GA to guide its adaptation.

Ronald et al. has also defined a qualitative emergence test based on two different viewpoints and a surprise factor [Ronald et al., 1999]. Specifically, from a designer's point of view, a design language $\mathcal{L}_1$ is used to describe local interactions. From an observer's point of view, an observed behavior is described by an observation language $\mathcal{L}_2$. Surprise is defined as the difference between the expected outcome of $\mathcal{L}_1$ as perceived by an observer and the observation by an observer using $\mathcal{L}_2$.

The complexity of AOC can have a direct implication on whether or not a problem is solvable. Standish has pointed out that the complexity of a complex system is context dependent and can be measured by the entropy of the system [Standish, 2001]. The entropy is calculated from the length of the system description, the size of the alphabet needed to encode the description, and the size of all equivalent descriptions. Standish has applied this principle to measure the complexity of the artificial life system, Tierra [Standish, 1999]. Nehaniv and Rhodes have also defined a hierarchical complexity measure for a biological system [Nehaniv and Rhodes, 2000]. They have defined a maximal complexity measure as the least number of active computing levels that are required to hierarchically build a finite transformation semigroup from simple components [Nehaniv and Rhodes, 2000]. It increases as the computational power increases, but in a bounded manner.

The evolvability of AOC [Nehaniv, 2000c] refers to its ability to evolve an optimal solution. Zhang and Shimohara have made an interesting observation in their experiments with Tierra in which the domination of certain genotypes does not mean that evolution stops [Zhang and Shimohara, 2000]. A new species comes to dominate a population beyond a seemingly stagnated state, if the population is allowed to continue to evolve. They have defined an index to measure evolutionary actions over time. This index is a weighted entropy of size distribution of a Tierran organism. The weight function is set to be the ratio of Tierran sizes at adjacent steps. The experimental results show that during the period of strong evolutionary actions, the entropy increases. The entropy will then drop to a low level and stay there when evolutionary actions dampen. Nehaniv has proposed to measure evolvability as the rate of complexity (defined above) increase [Nehaniv, 2000b]. By considering complexity in a longer time, Nehaniv has shown that the proposed evolvability measure is upper-bounded by one or one plus the complexity of an individual's ancestor. Here, the specific upper bound depends on the type of step that has occurred during the evolution from the said ancestor to the said individual.

3.7.   Simulation Environments

Performing experiments in AOC requires writing either a simulation environment or tailor-made programs. This section reviews two publicly available and widely used environments.

StarLogo [Resnick, 1994, StarLogo, 2000] is a simulation environment for explorations of colony-like behavior. It is a parallel implementation of the Logo programming language. A StarLogo world mainly consists of two classes of objects: environment and creature. An environment is a grid of patches that are dynamical and perform functions like diffusing pheromone. The patches can be inhabited by autonomous creatures, called turtles, or other species named by the programmer. Each turtle as well as patch can be programmed with certain behavior. The status of any patch and turtle can be queried and the objects related to patches can be altered by any turtle. Status update of all objects in StarLogo is synchronous. The latest version of StarLogo comes with a visualizer where a user can visually design the world of simulations. Many simulations have been implemented, such as termites, slime molds, traffic jam, among others. It is a good starting point to experiment with AOC. However, computational systems have not yet been reported.

Swarm [Minar et al., 1996, Swarm, 1994] is a larger scale simulation environment where users can perform simulations of complex adaptive systems. The basic element is a swarm, which is a collection of agents with a schedule of events. Swarms can consist of other swarms. This is similar to the recursive relationship between emergent autonomy and synthetic autonomy in AOC. Everything in Swarm, including the environment, is an agent with specific behavior. This provides a high degree of flexibility for engineers to experiment with different setups. Given in the object oriented simulation system is a set of libraries, namely, simulation libraries, software support libraries, and model-specific libraries. These libraries of object classes make the modeling process easier by hiding certain simulation-specific technicalities, such as an order of action execution and visualization, allowing modelers to concentrate on the problem-specific issues, such as agent behaviors and events of a particular simulation.

3.8.   Summary

In this chapter, we have identified and discussed several important issues in designing and engineering an AOC system. Generally speaking, an autonomous entity in an AOC system contains several common functional modules for sensing information, making decisions, and executing actions. Three phases are involved in developing AOC systems for achieving various objectives. The engineering steps in these phases can differ from one system to another in terms of human involvement, depending on our knowledge about problems or systems at hand. Moreover, we have identified several key features and attributes in an engineered AOC system, such as locality and amplification. We have discussed some important performance characteristics as well as determining factors in its performance.

Exercises

3.1 Define the behaviors of team sports players, as in basketball, American football, or soccer, using the autonomy modeling process discussed in this chapter.

3.2 What difference would it make if the behavior of an autonomous entity is totally deterministic?

3.3 Direct communication between players, by spoken or body language, is important to the success of a team in team sports. However, this is not explicitly defined in the autonomy modeling process. Can you think of any practical reasons why? How would you model direct communication?

3.4 Apart from the performance measurements mentioned in this chapter, can you think of any other useful metrics?

3.5 How would an AOC algorithm benefit from parallel machines or cluster computing resources? How does the ability to perform synchronous status updates affect the performance of an AOC algorithm?

Chapter 4

A Formal Framework of AOC

4.1.　Introduction

The preceding chapter has presented an overview of autonomous entities in an AOC system as well as the major phases in developing an AOC system. In essence, AOC systems implemented with different approaches have similar structures and operating mechanisms. In order to better describe and model AOC systems using a unified language, in this chapter we will provide a formal, common framework for AOC systems [Liu et al., 2004a]. In particular, we will formally introduce such key elements as environment, entity, and interaction. The formal definitions are meant to show the inter-relationships among common concepts involved in AOC. Based on the definitions, we will highlight the essence of AOC systems, i.e., the process of self-organization in autonomous entities.

4.2.　Elements of an AOC System

An AOC system usually contains a group of autonomous entities and an environment where entities reside. We can formally define an AOC system as follows:

Definition 4.1 (Autonomy oriented computing system) *An AOC system is a tuple* $\langle \{e_1, e_2, \cdots, e_i, \cdots, e_N\}, \mathbf{E}, \Phi \rangle$, *where* $\{e_1, e_2, \cdots, e_i, \cdots, e_N\}$ *is a group of autonomous entities;* $\mathbf{E}$ *is an environment in which entities reside;* Φ *is a system objective function, which is usually a nonlinear function of entity states.*

4.2.1   Environment

As we have mentioned in Section 3.2, the environment of an AOC system plays three roles. As compared with the other two roles, the second one, i.e., a communication medium among autonomous entities, is more directly related to the behavior, interaction, and self-organization of entities. Specifically, an environment plays its communication medium role through its state changes as caused by the primitive behavior of autonomous entities. For this role, we can formally define an environment as follows:

Definition 4.2 (Environment) *Environment* $\mathbf{E}$ *is characterized by a set* $\mathcal{ES} = \{es_1, es_2, \cdots, es_i, \cdots, es_{N_{\mathcal{ES}}}\}$, *where each* $es_i \in D_{es_i}$ *corresponds to a static or dynamical attribute,* $N_{\mathcal{ES}}$ *denotes the number of attributes. At each moment,* $\mathcal{ES}$ *represents the current state of environment* $\mathbf{E}$. *Thus, the state space of* $\mathbf{E}$ *is given by* $D_{\mathcal{ES}} = D_{es_1} \times D_{es_2} \times \cdots \times D_{es_i} \times \cdots \times D_{es_{N_{\mathcal{ES}}}}$.

For example, in the AOC-based image segmentation shown in Section 2.3, an environment is characterized by a static attribute, i.e., the grey-scale intensity corresponding to each pixel.

4.2.2   Autonomous Entities

As the basic elements of an AOC system, autonomous entities achieve their respective goals by performing their primitive behaviors and complying with their behavioral rules. Through interactions, entities can self-organize them in order to achieve the system goal of problem solving or system modeling. Here, we define an autonomous entity as follows:

Definition 4.3 (Autonomous entity) *An autonomous entity* $\mathbf{e}$ *is a tuple* $\langle \mathcal{S}, \mathcal{F}, \mathcal{G}, \mathcal{B}, \mathcal{R} \rangle$, *where* $\mathcal{S}$ *describes the current state of entity* $\mathbf{e}$. $\mathcal{F}$ *is an evaluation function.* $\mathcal{G}$ *is the goal set of entity* $\mathbf{e}$. $\mathcal{B}$ *and* $\mathcal{R}$ *are primitive behaviors and behavioral rules, respectively.*

Based on the differences in $\mathcal{S}, \mathcal{F}, \mathcal{G}, \mathcal{B}$, and $\mathcal{R}$, entities in an AOC system can be categorized into different classes. Before further description, let us define the neighbors of an entity.

Definition 4.4 (Neighbors) *The neighbors of entity* $\mathbf{e}$ *are a group of entities* $\mathcal{L}^e = \{l_1^e, l_2^e, \cdots, l_i^e, \cdots, l_{N_{\mathcal{L}}}^e\}$, *where* $N_{\mathcal{L}}$ *is the number of neighbors. The relationship (e.g., distance) between each neighbor* l_i^e *and entity* $\mathbf{e}$ *satisfies certain application-dependent constraint(s).*

In different AOC systems, the neighbors of an entity can be fixed or dynamically changed.

At each moment, an entity is in a certain state. It, according to its behavioral rules, selects and performs its primitive behavior in order to achieve certain goals with respect to its state. While doing so, it needs to interact with its neighbors or its local environment to get necessary information. In the following, we will further describe the notions of state, evaluation function, goal, primitive behavior, and behavioral rule for an autonomous entity.

Definition 4.5 (State) *State S of autonomous entity **e** is characterized by a set of static or dynamical attributes, i.e., $S = \{s_1, s_2, \cdots, s_i, \cdots, s_{N_S}\}$, where $s_i \in D_{s_i}$ and N_S denotes the number of attributes. Thus, $D_S = D_{s_1} \times D_{s_2} \times \cdots \times D_{s_i} \times \cdots \times D_{s_{N_S}}$ corresponds to the state space of entity **e**.*

As we have noted that in the AOC-based search (see Section 2.3), an autonomous entity is characterized by three dynamical attributes, i.e., its position and age, as well as a flag for indicating whether or not it is active.

Before an entity fires its behavioral rules to select its primitive behavior, it needs to assess its current condition, including its own internal state and/or those of its neighbors and environment. In some applications, while selecting its behavior, an entity needs to assess its choices, i.e., possible states at the next step.

Definition 4.6 (Evaluation function) *Autonomous entity **e** assesses its conditions using one of the following evaluation functions:*

- *Internal state:*

$$\mathcal{F} : \hat{D}_S \longrightarrow R, \tag{4.1}$$

- *State of environment:*

$$\mathcal{F} : \hat{D}_{\mathcal{E}S} \longrightarrow R, \tag{4.2}$$

- *States of neighbors:*

$$\mathcal{F} : \prod_{l_i^e \in \mathcal{L}^e} (\hat{D}_{S^{l_i^e}}) \longrightarrow R, \tag{4.3}$$

- *Internal state and that of environment:*

$$\mathcal{F} : \hat{D}_S \times \hat{D}_{\mathcal{E}S} \longrightarrow R, \; or \tag{4.4}$$

- *Internal state and those of neighbors:*

$$\mathcal{F} : \hat{D}_S \times \prod_{l_i^e \in \mathcal{L}^e} (\hat{D}_{S^{l_i^e}}) \longrightarrow R, \tag{4.5}$$

where R denotes the range of function $\mathcal{F}$ (e.g., the set of real numbers or integers). $\hat{D}_S$ is a Cartesian product of elements in a subset of $\{D_{s_i}\}$, i.e.,

$\hat{D}_S \subseteq D_S$. *Similarly,* $\hat{D}_{\mathcal{ES}}$ *is a Cartesian product,* $\hat{D}_{\mathcal{ES}} \subseteq D_{\mathcal{ES}}$. l_i^e *denotes the ith neighbor of entity* e. $\mathcal{S}^{l_i^e}$ *and* $D_{\mathcal{S}^{l_i^e}}$ *denote the current state and the state space of entity* l_i^e, *respectively.* $\hat{D}_{\mathcal{S}^{l_i^e}} \subseteq D_{\mathcal{S}^{l_i^e}}$. $\prod$ *is the Cartesian product operator.*

Note that in the above definition, we use $\hat{D}_S$, $\hat{D}_{\mathcal{ES}}$, and $\hat{D}_{\mathcal{S}^{l_i^e}}$, instead of D_S, $D_{\mathcal{ES}}$, and $D_{\mathcal{S}^{l_i^e}}$. This is because evaluation function $\mathcal{F}$ is based on a subset of attributes, which represents an autonomous entity as well as its neighbors and environment.

Generally speaking, the primitive behavior of entities in AOC systems is goal directed. The goal of an entity is defined as follows:

Definition 4.7 (Goal) *An entity,* e, *can be engaged in a set of goals over time, as denoted by* $\mathcal{G} = \{g_1, g_2, \cdots, g_i, \cdots, g_{N_G}\}$, *where* N_G *denotes the number of goals. Each goal* g_i *is to achieve a certain state* $\mathcal{S}'$ *such that evaluation function* $\mathcal{F}$ *takes a certain predefined value* α, *i.e.,* $g_i = \{\mathcal{S}'|\mathcal{F}(\cdot) = \alpha\}$, *where* α *is a constant.*

In an AOC system, at a given moment each entity e usually has only one goal and all entities may share the same goal. Although the primitive behavior of autonomous entities is goal directed, entities do not explicitly know the global goal of the whole system and the task that the system is performing.

Definition 4.8 (Primitive behavior) *An entity,* e, *can perform a set of primitive behaviors,* $\mathcal{B} = \{b_1, b_2, \cdots, b_i, \cdots, b_{N_B}\}$, *where* N_B *denotes the number of primitive behaviors. Each primitive behavior* b_i *is a mapping in one of the following forms:*

- *Self-reproduce:*

$$b_i : \mathbf{e} \longrightarrow \mathbf{e}^m, \qquad (4.6)$$

   which is a reproduction-like behavior. It denotes that entity e *replicates itself* m *times (i.e., breed* m *offspring);*

- *Die:*

$$b_i : \mathbf{e} \longrightarrow \emptyset, \qquad (4.7)$$

   which denotes that entity e *vanishes from the environment;*

- *Change internal state:*

$$b_i : \hat{D}_S \longrightarrow \hat{D}_S, \qquad (4.8)$$

- *Change state of environment:*

$$b_i : \hat{D}_{\mathcal{ES}} \longrightarrow \hat{D}_{\mathcal{ES}}, \qquad (4.9)$$

- *Change internal state and that of environment:*

$$b_i : \hat{D}_S \times \hat{D}_{\mathcal{E}S} \longrightarrow \hat{D}_S \times \hat{D}_{\mathcal{E}S}, \text{ or} \qquad (4.10)$$

- *Change internal state and those of neighbors:*

$$b_i : \hat{D}_S \times \prod_{l_i^e \in \mathcal{L}^e} (\hat{D}_{S^{l_i^e}}) \longrightarrow \hat{D}_S \times \prod_{l_i^e \in \mathcal{L}^e} (\hat{D}_{S^{l_i^e}}), \qquad (4.11)$$

where $\hat{D}_S$, $\hat{D}_{\mathcal{E}S}$, $\hat{D}_{S^{l_i^e}}$, and $\prod$ have the same meanings as those in Definition 4.6.

Definition 4.9 (Behavioral rule) *The behavioral rule set for entity e is $\mathcal{R} = \{r_1, r_2, \cdots, r_i, \cdots, r_{N_{\mathcal{R}}}\}$, where $N_{\mathcal{R}}$ denotes the number of rules. Each behavioral rule r_i is to select one or more primitive behaviors to perform. Behavioral rules can be classified into two types:*

- *Evaluation-based rules:*

$$r_i : Ran(\mathcal{F}) \longrightarrow \{\hat{\mathcal{B}}\}, \qquad (4.12)$$

where $Ran(\mathcal{F})$ denotes the range of evaluation function $\mathcal{F}$. $\hat{\mathcal{B}} \subseteq \mathcal{B}$. $\{\hat{\mathcal{B}}\} \subseteq 2^{\mathcal{B}}$.

- *Probability-based rules:*

$$r_i : [0,1] \longrightarrow \{\hat{\mathcal{B}}\}, \qquad (4.13)$$

where each subset $\hat{\mathcal{B}}$ is assigned a probability, $p_{\hat{\mathcal{B}}}$, which may be fixed or dynamically changed over time. This type of rule probabilistically selects a set of primitive behaviors from $\mathcal{B}$.

For example, in the AOC-based feature search and extraction, an autonomous entity has four primitive behaviors, namely, breeding, pixel labeling, diffusion, and decay. At each step, it chooses to breed some offspring and label its current position, or diffuse to another position and decay, based on the assessment of its neighboring region. Here, the behavioral rule is evaluation-based.

4.2.3 System Objective Function

As a global measurement for the performance of an AOC system, system objective function Φ guides the system to evolve towards certain desired states or patterns.

Definition 4.10 (System objective function) *In an AOC system, system objective function Φ is defined as a function of the states of some entities. In different applications, Φ can be categorized into two types. Let $\{e_i\}$ be a group of entities.*

- *State-oriented function:*

$$\Phi : \prod_{e_k \in \{e_i\}} \hat{D}_{S^{e_k}} \longrightarrow R^m, \qquad (4.14)$$

where $\hat{D}_{S^{e_k}}$ is a subset of the state space, $D_{S^{e_k}}$, of entity e_k, R is the set of real numbers or integers, and m denotes the dimensionality of the space of Φ.

- *Process-oriented function:*

$$\Phi : \left\{ \prod_{e_k \in \{e_i\}} \hat{D}_{S^{e_k}} \right\}^{\Pi} \longrightarrow R^m, \qquad (4.15)$$

where $\{\cdot\}^{\Pi}$ denotes a series of multiplications of the elements inside the braces.

It should be pointed out that Φ is usually a nonlinear function. In the above definition, the first type of Φ concerns the state of an AOC system, which is measured with an m-dimensional vector. Specifically, the system aims at certain desired states. With the second type of Φ, an AOC system is intended to exhibit certain desired characteristics or patterns in its evolutionary process.

4.3. Interactions in an AOC System

The emergent behavior of an AOC system originates from the interactions among the elements of the system. This section addresses the interactions in an AOC system. Generally speaking, there are two kinds of interaction, namely, (1) interactions between autonomous entities and their environment and (2) interactions among entities.

4.3.1 Interactions between Entities and their Environment

The interactions between an autonomous entity and its environment can be described through the state changes in the environment, as caused by the primitive behavior of the entity.

Definition 4.11 (Interactions between an entity and its environment) *The interactions between entity* **e** *and its environment* **E** *are modeled as a sequence of mappings* $\{\mathcal{I}_{eE}\}$, *where* $\mathcal{I}_{eE}$ *has one of the following forms:*

$$\mathcal{I}_{eE} : \hat{D}_{\mathcal{E}S} \longrightarrow_{b_i} \hat{D}_{\mathcal{E}S}, \ or \qquad (4.16)$$

$$\mathcal{I}_{eE} : \hat{D}_S \times \hat{D}_{\mathcal{E}S} \longrightarrow_{b_i} \hat{D}_S \times \hat{D}_{\mathcal{E}S}, \qquad (4.17)$$

where '$\longrightarrow_{b_i}$' indicates that $\mathcal{I}_{eE}$ *is in fact a primitive behavior, b_i, of entity* **e**, *by performing which entity* **e** *can change the state of its environment (see Definition 4.8, Equations 4.9 and 4.10).*

Figure 4.1 presents a schematic diagram of the interactions between two entities e_A and e_B and their environment **E**.

4.3.2 Interactions among Entities

Different AOC systems may have different fashions of interactions among their entities. The interactions among entities can be categorized into two types: direct and indirect. Which type of interaction will be used in an AOC system is determined by specific applications.

Direct interactions are implemented through either direct state information exchanges among entities, or direct effects of the primitive behavior of one entity on the states of others. In an AOC system with direct interactions, each entity can interact with its neighbors. Figure 4.2 presents a schematic diagram of the direct interactions between two entities e_A and e_B.

Definition 4.12 (Direct interactions among entities) *The direct interactions between entities* e_A *and* e_B *are modeled as a sequence of mapping tuples* $\{\langle \mathcal{I}_{AB}, \mathcal{I}_{BA} \rangle\}$,

$$\mathcal{I}_{AB} : \hat{D}_{SA} \times \hat{D}_{SB} \longrightarrow_{b_A} \hat{D}_{SA} \times \hat{D}_{SB}, \qquad (4.18)$$

and

$$\mathcal{I}_{BA} : \hat{D}_{SB} \times \hat{D}_{SA} \longrightarrow_{b_B} \hat{D}_{SB} \times \hat{D}_{SA}, \qquad (4.19)$$

where '$\longrightarrow_{b_A}$' and '$\longrightarrow_{b_B}$' denote that $\mathcal{I}_{AB}$ *and* $\mathcal{I}_{BA}$ *are two primitive behaviors of entities* e_A *and* e_B, *respectively, which are related to the states of neighboring entities. Here, if we take* e_A *and* e_B *as neighbors of each other, we can regard that '$\longrightarrow_{b_A}$' and '$\longrightarrow_{b_B}$' are originated from Equation 4.11 in Definition 4.8.* $\hat{D}_{SA}$ *and* $\hat{D}_{SB}$ *are subsets of the state spaces of entities* e_A *and* e_B, *respectively.*

Indirect interactions are carried out through the environment of entities that functions as a communication medium. There are two phases involved in such

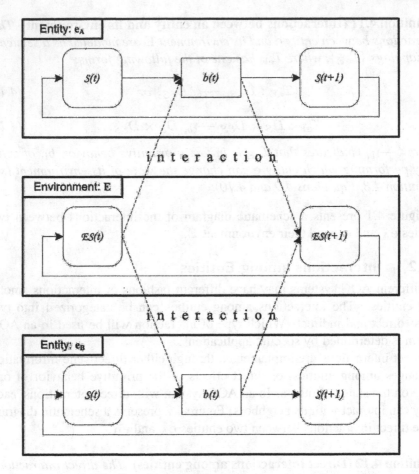

Figure 4.1. Interactions (i.e., the dashed lines) between two entities e_A and e_B and their environment E, which are caused by primitive behavior (i.e., $b(t)$). The solid lines denote the state changes in entities and their environment.

interactions: (1) Through the interactions between an entity and its environment, the entity 'transfers' its information to the environment; (2) Other entities (e.g., local neighbors) will sense and evaluate such information, and act accordingly.

Definition 4.13 (Indirect interactions among entities) *The indirect interactions between entities* e_A *and* e_B *are modeled as a sequence of mapping tuples* $\{\langle \mathcal{I}_{AE}, \mathcal{I}_{BE} \rangle\}$, *where interaction* $\mathcal{I}_{AE}$ *between entity* e_A *and environment* E *occurs before interaction* $\mathcal{I}_{BE}$ *between entity* e_B *and environment* E. *That is,*

- *At time t:*

$$\mathcal{I}_{AE} : \hat{D}_{\mathcal{ES}} \longrightarrow_{b^A} \hat{D}_{\mathcal{ES}}, \; or \qquad (4.20)$$

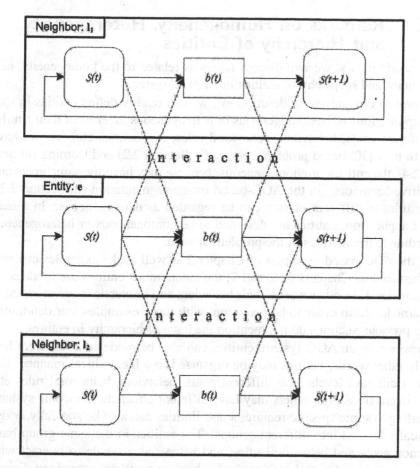

Figure 4.2. Direct interactions (i.e., the dashed lines) between entity **e** and its two neighbors, namely, entities **l₁** and **l₂**. The solid lines denote the state changes in the three entities as caused by their primitive behaviors.

$$\mathcal{I}_{AE} : \hat{D}_{SA} \times \hat{D}_{ES} \longrightarrow_{bA} \hat{D}_{SA} \times \hat{D}_{ES}; \qquad (4.21)$$

■ *At time t′:*

$$\mathcal{I}_{BE} : \hat{D}_{ES} \longrightarrow_{bB} \hat{D}_{ES}, \ or \qquad (4.22)$$

$$\mathcal{I}_{BE} : \hat{D}_{SB} \times \hat{D}_{ES} \longrightarrow_{bB} \hat{D}_{SB} \times \hat{D}_{ES}, \qquad (4.23)$$

where $t < t'$; ' $\longrightarrow_{bA}$ ' and ' $\longrightarrow_{bB}$ ' denote that $\mathcal{I}_{AB}$ and $\mathcal{I}_{BA}$ are two primitive behaviors of entities e_A and e_B, respectively; $\hat{D}_{SA} \subseteq D_{SA}$ and $\hat{D}_{SB} \subseteq D_{SB}$, where D_{SA} and D_{SB} are the state spaces of entities e_A and e_B, respectively.

4.4.   Remarks on Homogeneity, Heterogeneity, and Hierarchy of Entities

In what follows, we will discuss issues as related to the homogeneity, heterogeneity, and hierarchy of entities in an AOC system.

Based on our preceding descriptions, we can readily define entities in specific applications to be homogeneous or heterogeneous in terms of their similar or different goals, behaviors, behavioral rules, interactions etc.. As we have seen in the AOC-based problem solving (in Section 2.2) and learning (in Section 2.4), the entities are homogeneous, because they have the same goals and primitive behaviors. In the AOC-based image segmentation (in Section 2.3), the entities in different classes can be regarded as heterogeneous. In a real-world application, entities are designed to be homogeneous or heterogeneous according to the features of the problem at hand.

In the AOC-based examples of Chapter 2 as well as the examples that will be illustrated in Chapters 5, 6, and 7, the autonomous entities are modeled at the same level. In other words, their behaviors and interactions are modeled at the same level. In order to be consistent with those examples, our definitions in the previous sections do not mention the issue of hierarchy in entities.

However, in an AOC system, entities can also be modeled at different levels. In other words, entities may be organized in a hierarchical manner. Entities at different levels have different goals, behaviors, behavioral rules etc. Their interactions with others may have different effects to the whole system, according to some specific requirements. Entities can also be statically or dynamically formed into different groups. The entities in the same group have common goals and behavioral rules, and behave as a whole to interact with other groups or individual entities. In this book, we will not extend our discussions to this case.

4.5.   Self-Organization in AOC

In an AOC system, self-organization plays a crucial role. In this section, we will introduce the notion of self-organization and show how self-organization is utilized to achieve the goal of an AOC system.

4.5.1   What is Self-Organization?

The term "self-organization" was first introduced by Ashby in 1947 [Ashby, 1947, Ashby, 1966]. The phenomenon of self-organization exists in a variety of natural systems, such as galaxies, planets, compounds, cells, organisms, stock markets, and societies [Bak, 1996, Ünsal, 1993, Lucas, 1997]. It is involved in many disciplines, including biology, chemistry, computer science, geology, sociology, and economics [Liu, 2001, Bak, 1996, Kauffman, 1993]. Several theories have been developed to describe self-organizing systems. They include the dissipative structure theory of Prigogine [Nicolis and Prigogine, 1977, Prigogine, 1980] and the synergetics theory of Haken [Haken, 1983a, Haken, 1983b, Haken, 1988].

Generally speaking, a self-organizing system consists of two main elements: entities and an environment where the entities are situated. In a self-organizing system, entities are autonomous and behave rationally according to their behavioral rules. There is no explicit external control on the system, i.e., it is not governed by external rules.

The behavior of entities in a self-organizing system can be generalized into three steps [Ünsal, 1993]. First, entities sense the environment or receive signals from other entities. Secondly, based on the information received, entities make rational decisions on what to do next. Finally, entities behave according to their decisions. Their behavior will in turn affect the environment and the behavior of other entities. By following these three steps, an entity carries out its interaction with its environment or other entities.

The essence of a self-organizing system lies in interactions among its entities and the environment [Kauffman, 1993, Nicolis and Prigogine, 1977, Prigogine, 1980]. Through interactions, a self-organizing system can aggregate and amplify the outcome of entity behavior. Consequently, it eventually exhibits certain emergent behaviors or patterns.

An emergent behavior or pattern in a self-organizing system may correspond to a solution of a certain problem. By virtue of self-organization, researchers have proposed some methods for solving practical problems. Introducing the idea of self-organization into neural networks is a successful example. Self-organization methods have also been demonstrated in various applications, such as image feature extraction as described in [Liu et al., 1997]. In the following section, we will see that self-organization is the fundamental mechanism of an AOC system.

For more details on self-organization and self-organizing systems, readers are referred to [Liu, 2001, Bak, 1996, Ashby, 1966, Haken, 1983a, Haken, 1983b, Haken, 1988, Kauffman, 1993, Nicolis and Prigogine, 1977, Prigogine, 1980].

4.5.2 How Does an AOC System Self-Organize?

How does an AOC system self-organizes so that it can solve some hard computational problems (e.g., n-queen problems, image segmentation, and global optimization) and exhibit complex behavior (e.g., the surfing behavior of users on the Internet)? The key lies in the self-organization of its autonomous entities, from which all emergent behaviors originate.

In order to achieve their respective goals, individual entities autonomously make decisions on selecting or performing their primitive behavior. While selecting or performing primitive behavior, they need to consider not only the information of their own states, but also that of their neighbors and their environment. To do so, they will either directly interact with each other, or indirectly interact via their environment, to exchange their state information or affect the states of each other. By performing behaviors, entities change their states towards their respective goals. Because autonomous entities take into consideration the states of others while behaving, from a global point of view, entities aggregate and amplify the outcome of their behaviors in order to achieve the global goal of the system.

Let us take the AOC-based search (see Section 2.3) as an example to demonstrate the process of self-organization. In this example, an entity tries to find a pixel that belongs to a certain homogeneous region. At such a pixel, its evaluation value will be better than those at other pixels. On one hand, when an entity finds a desired pixel, it will reproduce some offspring within its local environment, where the offspring will most probably find other desired pixels. This mechanism acts as positive feedback, through which the AOC system aggregates and amplifies the successful behavioral outcome of entities. On the other hand, if an entity cannot find a desired pixel after predefined steps, i.e., its lifespan, it will be deactivated. Through this positive feedback mechanism, the AOC system eliminates those entities with poor performance. From a global point of view, we can note that at the beginning steps, few entities can successfully find the desired homogeneous region. As the search progresses, more entities which are either reproduced or diffusing are able to locate pixels of a homogeneous region that has been found. This nonlinear process will continue until it achieves a state where all active entities stay in a certain homogeneous region. Hence, the whole AOC system is globally optimized.

Figure 4.3 shows a schematic diagram of the process of self-organization in an AOC system. In general, we can define the process of self-organization as follows:

Definition 4.14 (Self-organization of entities) *The process of self-organization of entities* $\{e_i\}$ *in an AOC system is a sequence(s) of state transitions* $\{\{S_t^{e_i}\}|t = 0, \cdots, T\}$*, which is subject to the following two constraints[1]:*

1. Locally, for each entity e_i*,*

$$Pr\left(\mathcal{F}(S_{t+1}^{e_i}) - \mathcal{F}(S_t^{e_i}) \succ 0\right) > 0, \qquad (4.24)$$

where $\mathcal{F}(S_t^{e_i})$ *returns the evaluation value of entity* e_i*'s state at time* t*;* $Pr(\cdot)$ *returns a probability;* $\mathcal{F}(S_{t+1}^{e_i}) - \mathcal{F}(S_t^{e_i}) \succ 0$*, i.e.,* $\mathcal{F}(S_{t+1}^{e_i}) \succ \mathcal{F}(S_t^{e_i})$*, denotes that the new state of entity* e_i *at time* $t + 1$ *is 'better' (say, higher) than the one at time* t*.*

---

[1]The number of entities, i.e., $|\{e_i\}|$, during the process of self-organization may vary over time.

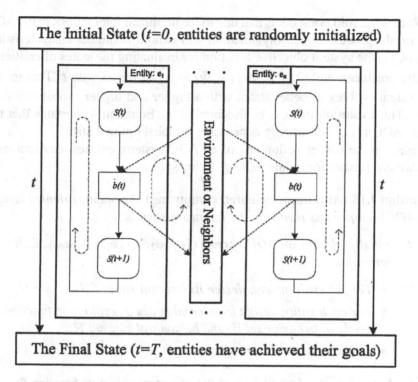

The Initial State (*t=0*, entities are randomly initialized)

The Final State (*t=T*, entities have achieved their goals)

Figure 4.3. The process of self-organization in an AOC system. The dash-dotted line denotes the outcome aggregation of entity behaviors on the states of entities and that of their environment.

2. *Globally, for the whole system,*

$$Pr\left(\Phi_{t+1} - \Phi_t \succ 0\right) > 0, \qquad (4.25)$$

where Φ_t *and* Φ_{t+1} *denote the value vectors of system objective function* Φ *at time* t *and* $t + 1$, *respectively.* $\Phi_{t+1} - \Phi_t \succ 0$ *means that the system evolves to a better state at time* $t + 1$.

Given the above local and global constraints, the AOC system eventually reaches a state where

$$\Phi_T = Opt(\Phi_t), \qquad (4.26)$$

where $Opt(\cdot)$ *returns an optimal value of the system objective function.*

In the above definition, the local constraint indicates that autonomous entities probabilistically evolve to better states at each step, although for some entities this may not hold. The global constraint reflects that the whole AOC system also probabilistically evolves to a better state in terms of its system objective function Φ at each step.

What we should point out is that due to the nonlinear interactions in the AOC system, the process of self-organization will exhibit nonlinear characteristics. That is, (1) the system objective function for evaluating the states of entities is usually nonlinear, and (2) $Pr\left(\Phi_{t+1} - \Phi_t \succ 0\right)$ increases with t. That means the system evolves to better states with a higher and higher probability over time. This evolution process is also nonlinear. Equation 4.26 states that the whole AOC system eventually approaches a global optimal state.

Based on the earlier definitions of an AOC system, entities, environments etc., we can further formulate AOC as follows:

Definition 4.15 (Autonomy oriented computing) *Autonomy oriented computing (AOC) is a process involving the following steps:*

Step 1. Initially, design an AOC system $\langle\{e_1, e_2, \cdots, e_i, \cdots, e_N\}, \mathbf{E}, \Phi\rangle$, in particular,

- *For the environment, design its internal states $\mathcal{ES}$;*
- *For each entity, design its internal states $\mathcal{S}$, evaluation function $\mathcal{F}$, goals $\mathcal{G}$, behavior set $\mathcal{B}$, and behavioral rule set $\mathcal{R}$;*
- *Design system objective function Φ.*

Step 2. Determine the desired 'value' Φ^* of system objective function Φ.

Step 3. Execute the designed AOC system, and then evaluate the current 'value' Φ'.

Step 4. Define an optimization function $\Psi = |\Phi' - \Phi^*|$ as a guidance for autonomy oriented computing.

Step 5. If Ψ is not optimized, modify or fine-tune the parameters of entities or the environment according to Ψ. In problem solving, 'optimized' means the system parameters are well set and the AOC system can successfully and efficiently find a solution to the problem. In system modeling, 'optimized' means the prototyping system can actually simulate the system to be modeled.

Step 6. Repeat the above steps until Ψ is optimized.

4.6. Summary

For an AOC system, autonomous entities and an environment are its key elements. Interactions between entities and their environment are the force that drives an AOC system to evolve towards certain desired states. Self-organization is the essential process of its working mechanism. In this chapter, we have formally defined the above notions as well as their intentions, and have provided a general framework for an AOC system. Based on the framework, we know how to build an AOC system, given a problem to be solved or a complex system to be modeled. Specifically, we have better knowledge of:

1. How to formally characterize autonomous entities?

2. How to design and characterize an environment based on a task at hand?

3. How to design the interactions between autonomous entities and their environment in order to facilitate the aggregation of behavioral effects of entities?

4. How to design the primitive behaviors and behavioral rules of entities in order to achieve the self-organization of entities and emerge certain desired states or patterns?

Exercises

4.1 In this chapter, autonomous entity e is defined as a 5-tuple, i.e., $e = \langle \mathcal{S}, \mathcal{F}, \mathcal{G}, \mathcal{B}, \mathcal{R} \rangle$. Explain whether or not these 5 parameters are sufficient to characterize an autonomous entity.

4.2 This chapter presents a probability-based definition of self-organization in an AOC system. Give another formal or mathematical definition of self-organization based on your understanding.

4.3 Chapter 2 has illustrated three examples on employing the ideas of AOC to tackle practical problems. For those examples, use the AOC framework presented in this chapter to:

(a) Formulate their autonomous entities and environments in detail;

(b) Identify their interaction types;

(c) Identify and mathematically formulate the specific self-organizing mechanisms used.

4.4 The AOC framework presented in this chapter is a general one. Take a step further to propose and develop specific frameworks for three AOC approaches by emphasizing their differences.

4.5 Based on the presented general framework, describe some crucial steps that can influence the process of self-organization towards desired collective behaviors.

4.6 As we have stated in Section 4.4, entities may be modeled in a hierarchical manner. Extend the AOC framework presented in this chapter to a case, in which entities are hierarchically organized.

Part II
AOC IN DEPTH

Chapter 5

AOC in Constraint Satisfaction

In Part I, we have described the key concepts, approaches, and framework of autonomy oriented computing (AOC). In the following three chapters, we will present and discuss three representative examples, which respectively correspond to the three AOC approaches as introduced earlier. By doing so, we aim to demonstrate how AOC can be engineered and evaluated in tackling practical problem solving and complex systems modeling tasks.

5.1. Introduction

In the real world, many challenging problems, such as e-learning in Section 5.1.1, are distributed in nature. Such problems cannot be solved in a centralized fashion. Instead, they require certain distributed computing mechanisms. AOC-by-fabrication attempts to provide a multi-entity based, distributed methodology for solving naturally distributed problems. In this chapter, we will take constraint satisfaction problems (CSPs) and satisfiability problems (SATs) as examples to demonstrate how the AOC-by-fabrication approach can be used. At the same time, we will highlight the key features of this approach.

The AOC-by-fabrication approach is designed by following some abstracted models of natural systems. Work in the field of artificial life (ALife) provides a basis for such an endeavor, as the definition of ALife indicates:

"The study of man-made systems that exhibit behavior characteristic of natural living systems" [Langton, 1989].

"...a field of study devoted to understanding life by attempting to abstract the fundamental dynamical principles underlying biological phenomena, and recreating these dynamics in other physical media, such as computers, making them accessible to new kinds of experimental manipulation and testing" [Langton, 1992].

Some well-known instances of ALife include visual arts [Sims, 1991], L-System [Prusinkiewicz et al., 1997, Prusinkiewicz and Lindenmayer, 1990] and Tierra [Ray, 1992]. These systems are intended to replicate some natural behaviors.

In this chapter, we will describe the steps in formulating and implementing the AOC-by-fabrication approach to solving computationally hard problems. Specifically, we will focus on a novel self-organization based method, called ERE, for solving constraint satisfaction problems (CSPs) [Kumar, 1992, Nadel, 1990] and satisfiability problems (SATs) [Folino et al., 2001, Gent and Walsh, 1993]. The ERE method involves a multi-entity system where each entity can only sense its local environment and apply a probability-based behavioral rule for governing its primitive behaviors, i.e., movements. The two-dimensional cellular environment records and updates the local values that are computed and affected according to the primitive behaviors (i.e., movements) of autonomous entities.

In solving a CSP or SAT with the ERE method, we first divide variables into several groups, and then represent each variable group with an entity whose possible positions correspond to the elements in the Cartesian product of variable domains. Next, we randomly place each entity onto one of its possible positions. Thereafter, the ERE system will keep on dispatching entities to choose their movements until an exact or approximate solution emerges. Experimental results on classical CSPs, i.e., n-queen problems, and some benchmark SAT testsets have shown that the ERE method can efficiently find exact solutions to n-queen problems, and can obtain comparable as well as stable performances in solving SATs. Particularly, it can find approximate solutions to both n-queen and SAT problems in just a few steps.

5.1.1   e-Learning

Many problems in Artificial Intelligence (AI) as well as in other areas of computer science and engineering can be formulated into CSPs or SATs. Some examples of such problems include: spatial and temporal planning, qualitative and symbolic reasoning, decision support, computational linguistics, scheduling, resource allocation and planning, graph problems, hardware design and verification, configuration, real time systems, robot planning, block world planning, circuit diagnosis, vision interpretation, and theorem proving. Here, we take e-learning as an example to show how a practical problem is translated into a CSP.

e-learning is a technology used to provide learning and training contents to learners via some electronic means, such as computers, Intranet, and Internet. Essentially, it bridges the minds of instructors and learners with IT technologies. Contents are usually structured into different modules, called learning objects, at different granularities, such as fragments (e.g., picture, figure, table, text), lessons, topics, units, courses, and curricula [Loser et al., 2002, IEEE, 2001, IEEE, 2002]. The reason for doing so is to increase the interoperability and scalability of an e-learning service environment [Loser et al., 2002]. Learning objects are located on different, usually geographically distributed, computers. Several smaller granular objects can constitute a bigger granular object. Generally speaking, there are two important types of relationship among learning objects [Loser et al., 2002]: content and ordering relations. A content relation describes the semantic interdependency among modules, while an ordering relation describes the sequence for accessing modules. In an e-learning service, these relations actually act as content constraints that should be satisfied during the service time.

In addition to the above mentioned content and ordering relations, an e-learning service can also involve system constraints that are related to the hardware and software of the service environment. Generally speaking, more than one learner can simultaneously access either the same or different contents. Learners are also geographically distributed. Although a learning object in an e-learning environment can be accessed by one or more learners each time, the number of learners is limited because of some specific factors, such as hardware bandwidth. Therefore, these factors play the roles of system constraints.

Example 5.1 *Figure 5.1 shows an illustrative e-learning service scenario, where four learners, A, B, C, and D, are accessing the e-learning service via clients connected to the service environment. The learners need to learn three knowledge modules, i.e., m_1, m_2, and m_3, provided by three geographically distributed computers. The constraints in the environment are:*

1. *Content constraint: To learn modules m_2 and m_3, module m_2 must be learned before module m_3.*

2. *System constraint: One module can serve at most two learners at any time.*

In order to successfully provide services to the above learners simultaneously, the distributed clients should collaborate with one another in order to determine a service sequence for each learner.

This collaborative service can be formulated as follows: For each learner $L \in \{A, B, C, D\}$, its client should coordinate with those of other learners in order to generate a combination $\{S_L^1, S_L^2, S_L^3\}$ of $\{1, 2, 3\}$ ($\forall i$, $S_L^i \in \{1, 2, 3\}$), which will be used as the sequence in which learner L accesses the knowledge modules. For example, if learner A is assigned a combination $\{2, 3, 1\}$, it means learner A should access the three modules in a sequence of 2, 3, and 1. The above mentioned constraints can be accordingly formulated as follows:

1. $\forall L \in \{A, B, C, D\}$, $S_L^i \in \{1, 2, 3\}$, and for $i \neq j$, $S_L^i \neq S_L^j$.

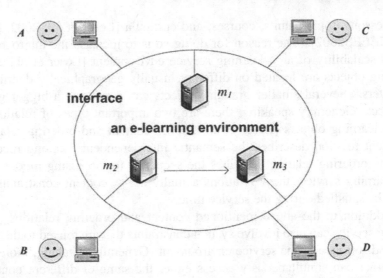

Figure 5.1. A scenario where an e-learning environment provides services to four learners simultaneously. All learners need to learn three modules, where module m_1 is independent, while modules m_2 and m_3 are dependent: Module m_2 has to be learned before module m_3. At any time, each module can only be accessed by at most two learners.

2. Content constraint: $\forall L \in \{A, B, C, D\}$, if $S_L^i = 2$ and $S_L^j = 3$, then it should guarantee $i < j$.

3. System constraint: $\forall i, j \in \{1, 2, 3\}$, $\sum_{L \in \{A,B,C,D\}} T(S_L^i = j) \leq 2$, where $T(\cdot)$ is a Boolean function to test whether or not an input proposition is true.

With the above three constraints, we have actually formulated the collaborative service into a distributed CSP, where distributed clients are responsible for assigning values to the variables of their learners.

5.1.2   Objectives

Constraint satisfaction problem (CSP) and satisfiability problem (SAT) are two types of classical NP-hard problem. In this section, we will provide the formal definitions of CSP and SAT.

Definition 5.1 (Constraint satisfaction) *A constraint satisfaction problem (CSP), P, consists of:*

1. A finite set of variables, $X = \{x_1, x_2, \cdots, x_i, \cdots, x_n\}$.

2. A domain set, containing a finite number of discrete domains for variables in X: $D = \{D_1, D_2, \cdots, D_i, \cdots, D_n\}$, $\forall i \in [1, n]$, $x_i \in D_i$.

3. A constraint set, $C = \{C(R_1), C(R_2), \cdots, C(R_i), \cdots, C(R_m)\}$, where each R_i is an ordered subset of the variables, and each constraint $C(R_i)$

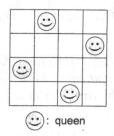

☺ : queen

Figure 5.2. A solution to a 4-queen problem.

is a set of tuples indicating the mutually consistent values of the variables in R_i.

Definition 5.2 (CSP Solution) *A solution, S, to a CSP is an assignment to all variables such that the assignment satisfies all given constraints. Specifically,*

1. *S is an ordered set, $S = \langle v_1, v_2, \cdots, v_i, \cdots, v_n \rangle$, $S \in D_1 \times D_2 \times \cdots D_i \times \cdots \times D_n$.*

2. *$\forall i \in [1, m]$, S_i is an ordered value set corresponding to variables in R_i, and $S_i \subseteq S$, $S_i \in C(R_i)$.*

Below is a classical CSP example.

Example 5.2 *An n-queen problem is a classical CSP. It is generally regarded as a good benchmark for testing algorithms and has attracted attention in the CSP community [Sosic and Gu, 1994]. This problem requires one to place n queens on an $n \times n$ chessboard, so that no two queens are in the same row, the same column, or the same diagonal. There exist solutions to n-queen problems with $n \geq 4$ [Bitner and Reingold, 1975, Sosic and Gu, 1994] (see Figure 5.2). The equivalent CSP can be stated as follows:*

$X = \{x_1, x_2, \cdots, x_i, \cdots, x_n\}$.

$D = \{D_1, D_2, \cdots, D_i, \cdots, D_n\}, \forall i, D_i = [1, n]$.

$C = \{C(R_u) | \forall i, j \in [1, n], C(R_u) = \{\langle b, c \rangle | b \in D_i, c \in D_j, b \neq c, i - j \neq b - c, i - j \neq c - b\}\}$.

Now, let us take a look at an SAT. Generally speaking, an SAT is to test whether there is (at least) one solution for a given propositional formula.

Definition 5.3 (Satisfiability problem) *A satisfiability problem (SAT), P, consists of:*

1. *A finite set of propositional variables, $X = \{x_1, x_2, \cdots, x_i, \cdots, x_n\}$.*

2. *A domain set, $D = \{D_1, D_2, \cdots, D_i, \cdots, D_n\}$, $\forall i \in [1, n]$, $x_i \in D_i$ and $D_i = \{True, False\}$.*

3. *A clause set, $CL = \{Cl(R_1), Cl(R_2), \cdots, Cl(R_i), \cdots, Cl(R_m)\}$, where each R_i is a subset of X, and each clause $Cl(R_i)$ is a disjunction of the literals corresponding to the variables in R_i.*

Definition 5.4 (SAT solution) *A solution, S, to an SAT is an assignment to all variables such that, under this assignment, the truth values of all given clauses are true, i.e.,*

1. *S is an ordered set, $S = \langle v_1, v_2, \cdots, v_i, \cdots, v_n \rangle$, $\forall i \in [1, n]$, $v_i \in \{True, False\}$, $S \in D_1 \times D_2 \times \cdots D_i \times \cdots \times D_n$.*

2. *$\forall i \in [1, m], T(Cl(R_i)) = True$, where $T(\cdot)$ is a function that returns the truth value of a clause.*

In [Gu, 1993], CSP and SAT were regarded as natural 'twins'. To some extent, we can consider SAT as a special type of CSP. In fact, for Definition 5.3, we can readily change item *3* to *3'*:

3'. A constraint set, $C = \{C(R_1), C(R_2), \cdots, C(R_i), \cdots, C(R_m)\}$, where each R_i is a subset of X, and each constraint $C(R_i)$ is a set of truth assignments to all variables in R_i, which satisfy $T(Cl(R_i)) = True$.

Thus, we can see that SAT is indeed a special case of CSP. Later in this chapter, we will take this point of view to treat SAT, for the convenience of description.

5.2.   Background

In this section, we will briefly survey conventional and self-organization based methods for solving CSPs and SATs. In addition, we will make a comparison between these methods and our ERE method.

5.2.1   Conventional Methods

Conventional methods for solving CSPs can be classified into generate-and-test (GT) methods and backtracking (BT) methods [Kumar, 1992]. A GT method generates each possible combination of variables systematically and then checks whether it satisfies all constraints, i.e., whether it is a solution. One limitation of this method is that it has to consider all elements of the Cartesian product of all the variable domains. In this respect, BT is more efficient than GT, as it assigns values to variables sequentially and then checks constraints for each variable assignment. If a partial assignment does not satisfy any of the constraints, it will backtrack to the most recently assigned variable and repeat the process again. Although this method eliminates a subspace from the

Cartesian product of all the variable domains, its computational complexity in solving most nontrivial problems remains to be exponential.

There has been some earlier work on how to improve the above mentioned BT method. In order to avoid thrashing in BT [Gaschnig, 1979, Kumar, 1992], consistency techniques (Arc Consistency and k-Consistency) have been developed by some researchers [Cooper, 1989, Han and Lee, 1988, Kumar, 1992, Mackworth, 1977, Mohr and Henderson, 1986]. These techniques are able to remove inconsistent values from the domains of variables. In order to avoid both thrashing and redundant work in BT [Kumar, 1992], a dependency directed scheme and its improvements have been proposed [Bruynooghe, 1981, Kumar, 1992, Rossi et al., 1990, Stallman and Sussman, 1977]. Other ways of increasing the efficiency of BT include the use of search orders for variables, values, and consistency checks. Nevertheless, even with such improvements, BT is still unable to solve nontrivial large-scale CSPs in a reasonable runtime.

It should be mentioned that there are also some research efforts on making GT smarter. The representatives of such efforts are stochastic and heuristic algorithms. Along this direction, one of the most popular ideas is to perform local search [Gu, 1992]. For large-scale n-queen CSPs, local search gives better results than a complete, or even incomplete, systematic BT method.

Methods for solving SATs cover two main branches: systematic search methods and local search methods [Hoos and Stützle, 1999]. Systematic search is a traditional way for solving an SAT that, like a BT method for a CSP, assigns values to partial variables and then checks whether there are some clauses unsatisfied. If this is the case, it will backtrack to a previous variable and assign it with another value, and then repeat this process. Otherwise, it will select a variable to branch until a solution is found or it finds that the problem is unsatisfiable. Such a method is complete because it guarantees to find out via search whether a given problem is satisfiable. Some examples of systematic search are POSIT, TABLEAU, GRASP, SATZ, and REL_SAT [Hoos and Stützle, 1999], all of which are based on the Davis-Putnam (DP) algorithm [Davis et al., 1962]. Algorithm 5.1 presents an outline of the Davis-Putnam algorithm.

Local search first appeared in 1992 when Selman, Levesque, and Gu almost simultaneously proposed the idea [Selman et al., 1992, Gu, 1992]. Generally speaking, local search starts with a complete and randomly initialized assignment, then checks whether it satisfies all clauses. If not, it will heuristically or randomly select a variable to flip (i.e., change its value). It repeats this process until a solution is found. Algorithm 5.2 presents an outline of a local search method.

As can be noted from Algorithm 5.2, a local search method contains three key concepts [Barták, 1998]:

Algorithm 5.1 The Davis-Putnam algorithm.

Input: SAT problem P (variable set X, clause set C), empty assignment S

Procedure *Davis-Putnam(P,S)*

    Given S, do unit propagation on P;

    if P is empty **then**

        Return S;

    end if

    if P contains empty clause **then**

        Backtrack;

    end if

    Choose a variable, x, from X;

    Davis-Putnam($P \cup v, S \cup v = True$);

    Davis-Putnam($P \cup \neg v, S \cup v = False$).

---

1. Configuration: one possible assignment to all variables, not required to be a solution;

2. Evaluation value: the number of satisfied constraints (in a CSP) or clauses (in an SAT);

3. Neighbor: a configuration obtained by flipping the assignment of a variable in the current configuration.

Although local search has been demonstrated to outperform systematic search, it is an incomplete method. That is to say, local search cannot prove that a given problem has no satisfying assignment. In addition, it cannot guarantee to find a solution for a satisfiable problem. Despite this, many improvements have been introduced to the local search method. There are two main streams: GSAT [Gu, 1992, Selman et al., 1992] and WalkSAT [Selman et al., 1994]. They all have many variants, such as GWSAT [Selman et al., 1994], GSAT/Tabu [Mazure et al., 1997, Steinmann et al., 1997], HSAT [Gent and Walsh, 1993], and HWSAT [Gent and Walsh, 1995] following GSAT, and WalkSAT/Tabu [McAllester et al., 1997], Novelty [McAllester et al., 1997], and R-Novelty [McAllester et al., 1997] following WalkSAT.

5.2.2 Self-Organization Based Methods

In the CSP literature, there exist two self-organization based methods. In [Kanada, 1992], Kanada provided a macroscopic model of a self-organizing system. In order to realize self-organizing computational systems, Kanada and Hirokawa further proposed a stochastic problem solving method based on local operations and local evaluation functions [Kanada and Hirokawa, 1994]. They also gave a computational model, called Chemical Casting Model (CCM), and

---

Algorithm 5.2 The local search algorithm.

Input: SAT problem P (variable set X, clause set C), assignment S
for $i = 1$ to *Max-Cycles* **do**
   Initialize assignment S;
   for $j = 1$ to *Max-Steps* **do**
     if $Evaluation(S) = |C|$ **then**
       Return S;
     else
       //Select a neighbor to move to;
       Choose a variable, x, from X;
       Flip the value of x in S;
     end if
   end for
end for
Return 'no solution found'.

---

applied it to some classical CSP problems, such as n-queen problems and graph coloring problems, to test their method. Another method was developed by Liu and Han, in which an artificial life model was implemented for solving large-scale n-queen problems [Liu and Han, 2001]. Interested readers may refer to Section 2.2.

Before we describe our AOC-based method, we should mention two related examples of self-organizing systems, cellular automata [Gutowitz, 1991, Liu et al., 1997] and Swarm [Swarm, 1994], that had some influence on our method. Cellular automata are dynamical systems that operate in discrete space and time. Each position in the space is called a cell and the state of the cell is locally specified according to a set of behavioral rules. All cells are updated synchronously. A cellular automaton self-organizes in discrete steps and exhibits emergent complex properties if certain behavioral rules are employed locally[Liu et al., 1997, Shanahan, 1994]. Swarm is a system for simulating distributed multi-entity systems. It involves three key components: a living environment, a group of entities with some behavioral rules, and a schedule for updating the environment and entities and for dispatching entity behaviors.

5.2.3   ERE vs. other Methods

As inspired by the previous models of cellular automata [Gutowitz, 1991, Liu et al., 1997], Swarm [Swarm, 1994], and the artificial life model in [Liu and Han, 2001], in this chapter we will present a new method for solving CSPs and SATs. The method is called ERE (Environment, Reactive rules, and Entities).

Like cellular automata and Swarm, an ERE system relies on self-organization. In ERE, each entity follows its local behavioral rules, and as a result, the system gradually evolves towards a solution state. From the point of view of solving a CSP, the ERE method may be regarded as an extended GT algorithm, somewhat like local search. However, the main differences between ERE and local search can be summarized as follows:

1. ERE evaluates a particular system state not by the number of dissatisfied constraints for the whole assignment as in local search, but by the number of dissatisfied constraints for every value combination of each variable group. The values of system state evaluation are stored in the environment of an ERE system.

2. Similar to cellular automata, entities in ERE can synchronously behave according to their behavioral rules, whereas local search is sequential.

3. In local search, the neighbors of an assignment are restricted to those that are different from the current assignment only in the value of one variable. In ERE, the neighbors of an assignment can be different in the values of several variables.

The following section will demonstrate that if there exists a consistent solution, the ERE system will be able to eventually find it. However, if there exists no exact solution that can satisfy all constraints, it will be able to generate an approximate one. Furthermore, our experiments will show that ERE is efficient in finding both exact and approximate solutions to a CSP in few steps. Generally speaking, it is more efficient than BT methods and more readily solves some classical CSPs than the local search algorithm.

5.3.   ERE Model

Problem solving is a domain with which many multi-agent applications are concerned. These applications are aimed at tackling computational problems in a distributed setting. In many cases, the problems to be solved are inherently distributed in nature [Ferber, 1999]. One way to formulate such problems is to treat them as distributed CSPs, as illustrated in Example 5.1.

Yokoo et al. have proposed several algorithms (i.e., asynchronous backtracking, asynchronous weak-commitment search, and multi-agent real-time-A* algorithm with selection) for solving distributed CSPs [Yokoo, 1995, Yokoo et al., 1998, Yokoo and Hirayama, 1998, Yokoo and Hirayama, 2000, Yokoo and Kitamura, 1996]. In these algorithms, agents are individual solvers for obtaining partial solutions. One typical feature of these algorithms is that all of them are complete. In other words, they can examine whether or not a problem at hand is solvable. If this is the case, they can find all solutions of the problem. Later, Silaghi et al. improved the work of Yokoo et al. by introducing a mechanism to check whether the message an agent receives is legal [Silaghi et al., 2001a, Silaghi et al., 2001b, Silaghi et al., 2001c].

In this chapter, we will introduce an AOC-based method for solving general CSPs (including distributed CSPs). In our method, the domain of a CSP or its variant is represented into a multi-entity environment. Thus, the problem of finding a solution to a CSP is reduced to that of how a group of entities find a certain desired state by performing their primitive behaviors in such an environment. Like the other AOC-based methods, ERE exhibits several unique characteristics as well as advantages in tackling problems that involve large-scale, highly distributed, locally interacting, and sometimes unreliable entities.

The notions of entity and multi-entity based ERE system can be defined as follows:

Definition 5.5 (ERE Entity) *An entity, a, in an ERE system is a virtual entity that represents one or more variables in a given problem. It essentially has the following abilities:*

1. *An ability to reside and behave (i.e., move around) according to a probability-based behavioral rule in a local environment as specified by the domains of the variable(s) it represents;*

2. *An ability to interact with its local environment;*

3. *An ability to be driven by certain goals.*

Definition 5.6 (ERE system) *An ERE system is a system that contains the following elements:*

1. *An environment, E, as specified by the solution space of a problem at hand, in which entities reside;*

2. *A set of reactive rules (including primitive behaviors and behavioral rules), R, governing the behaviors of entities and the interactions among entities and their environment. Essentially, they govern entities to assign values to their respective variables. In this sense, they are the laws of the entity universe;*

3. *A set of ERE entities, $E = \{a_1, a_2, \cdots, a_i, \cdots, a_n\}$, which represent all variable of the problem.*

This chapter will examine how exact or approximate solutions to CSPs can be self-organized by a multi-entity system consisting of E, R, and E. In other words, it will illustrate:

Environment + Reactive rules + Entities $\Longrightarrow$ Problem solving

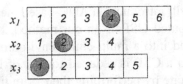

Figure 5.3.   An illustration of the entity model in Example 5.3.

5.3.1   General Ideas

An ERE system is intended to provide a distributed CSP solver. In ERE, we divide variables into groups. Each group contains one or more variables. A set of entities are employed to represent variable groups. The Cartesian products of the domains corresponding to all variable groups constitute an environment where entities reside. Therefore, each position of an entity indicates a value combination of the variables that it represents. An entity can move freely within a row and has its own primitive behaviors and behavioral rules. It tries to move to a position where the number of constraint violation is zero. We refer to such a position as zero-position (For more details, see Definition 5.7). The primitive behavior will locally determine how an entity moves and how the environment is updated. A solution state in ERE is reached when all entities (variable groups) can move to their zero-positions (consistent value combinations). In other words, a solution in ERE is specified by the positions of the distributed entities.

In the following paragraph, we will use two examples to illustrate how our ERE method works.

Example 5.3 *A CSP is given as follows:*

$X = \{x_1, x_2, x_3\}, n = 3.$

$D = \{D_1, D_2, D_3\}, D_1 = \{1, 2, 3, 4, 5, 6\}, D_2 = \{1, 2, 3, 4\},$
$\quad D_3 = \{1, 2, 3, 4, 5\}.$

$C = \{x_1 \neq x_2, x_1 > x_3\}.$

In ERE, the above example can be modeled as follows: We divide variables into three groups, i.e., each group contains only one variable. A lattice represents an environment, where each row represents the domain of a variable and the length of the row is equal to the size of the domain. In each row, there exists only one entity. In this case, the horizontal coordinate of an entity corresponds the value of a variable. As in Figure 5.3, there are three entities and they are all at zero-positions. The numbers encountered by these entities correspond to three values within the domains of the variables. Thus, Figure 5.3 corresponds to a solution state of $S = \langle 4, 2, 1 \rangle$.

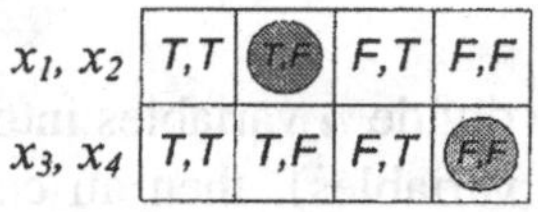

x_1, x_2	T,T	T,F	F,T	F,F
x_3, x_4	T,T	T,F	F,T	F,F

Figure 5.4. An illustration of the entity model in Example 5.4.

Example 5.4 *An SAT can be described as a special CSP as follows:*

$X = \{x_1, x_2, x_3, x_4\}$, $n = 4$.

$D = \{D_1, D_2, D_3, D_4\}$, $D_1 = D_2 = D_3 = D_4 = \{True, False\}$.

$C = \{$

$$
\begin{aligned}
T(x_1 \lor \neg x_2 \lor x_3) &= True, \\
T(x_1 \lor x_2 \lor \neg x_3) &= True, \\
T(x_2 \lor x_3 \lor \neg x_4) &= True, \\
T(\neg x_2 \lor \neg x_3 \lor x_4) &= True, \\
T(x_1 \lor x_3 \lor \neg x_4) &= True, \\
T(x_1 \lor x_3 \lor x_4) &= True, \\
T(\neg x_1 \lor x_2 \lor \neg x_3) &= True, \\
T(\neg x_1 \lor \neg x_2 \lor x_3) &= True, \\
T(\neg x_1 \lor \neg x_2 \lor \neg x_3) &= True
\end{aligned}
$$

$\}$.

Example 5.4 can be modeled in a similar way as Example 5.3. First, we divide the variables into two groups: $\{x_1, x_2\}$ and $\{x_3, x_4\}$. In this case, the Cartesian product of each variable group will be: $\{\{True, True\}, \{True, False\}, \{False, True\}, \{False, False\}\}$. Next, we use one entity to represent each variable group. The Cartesian product of the variable group will become the space for an entity to move around (see Figure 5.4). In Figure 5.4, there are two entities. Each entity will move within its respective row. The cells in a row correspond to the elements in the Cartesian product of a corresponding variable group.

In Figure 5.4, the two entities are at zero-positions. Therefore, the positions of entities in Figure 5.4 correspond to a solution state of $S = \langle True, False, False, False \rangle$ to the SAT problem in Example 5.4.

From the above examples, we can arrive at a general model of ERE in solving a CSP as follows:

CSP {X(variables), D(domains), C(constraints) } $\Rightarrow$ Multi-entity system;

D & C $\Rightarrow$ Cellular environment & Updating rules;

X $\Rightarrow$ Entities (each entity represents a group of variable(s));

Solution $\Rightarrow$ Positions of the entities.

5.3.2   Environment

Generally speaking, if we divide n variables into u groups (each group may have a different number of variables), then an environment, E, has u rows corresponding to the variable groups. For all $i \in [1, u]$, if we assume that row_i represents a variable group $\{x_{i1}, x_{i2}, \cdots, x_{ik}\}$, then it will have $|D_{i1} \times D_{i2} \times \cdots \times D_{ik}|$ cells. Each cell records two kinds of information: domain value and violation value.

Definition 5.7 (Environment) *An environment, E, in an ERE system can be defined as follows:*

1. Size

- $E = \langle row_1, row_2, \cdots, row_i, \cdots, row_u \rangle$.

- *u rows $\Leftrightarrow$ u variable groups.*

- *$\forall i \in [1, u]$, $row_i \Leftrightarrow$ all possible value combinations of variables in $\{x_{i1}, x_{i2}, \cdots, x_{ik}\} \Leftrightarrow D_{i1} \times D_{i2} \times \cdots \times D_{ik}$.*
  Hence, row_i has $|D_{i1} \times D_{i2} \times \cdots \times D_{ik}|$ cells. $row_i = \langle cell_{1i}, cell_{2i}, \cdots,$ $cell_{(|D_{i1} \times D_{i2} \times \cdots \times D_{ik}|)i} \rangle$. $e(j, i)$ refers to the position of $cell_{ji}$.

- *The size of E is $\sum_{i=1}^{u}(|D_{i1} \times D_{i2} \times \cdots \times D_{ik}|)$.*

2. Values

- **Domain value:** *$e(j, i).value$ records the jth value combination in $D_{i1} \times D_{i2} \times \cdots \times D_{ik}$.*

- **Violation value:** *$e(j, i).violation$ records in the current state how many constraints are unsatisfied, which are related to variables in position $e(j, i)$. $e(j, i).violation = m$ means that there are m constraints unsatisfied, which include some variable(s) in position $e(j, i)$. The violation value of a cell is dynamically updated, as entities keep on moving and their states are changing. After a movement by an entity, the violation values are updated by applying an updating-rule, which will be described in Section 5.3.3.*

- **Zero-position:** *a zero position (j, i) has $e(j, i).violation = 0$. That means all constraints, to which the variables of row_i are related, are satisfied.*

Now let us revisit Examples 5.3 and 5.4 based on the concepts of Definition 5.7.

Specifically, Figure 5.5(a) shows the position of an entity at $(1, 2)$. Figure 5.5(b) shows the domains of three variables. row_1 represents values in

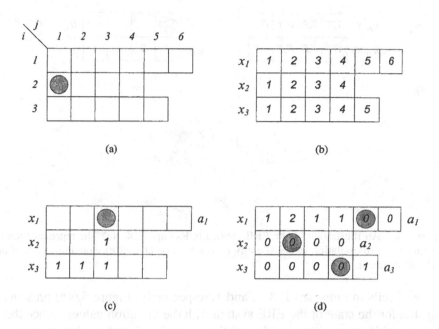

Figure 5.5. An illustration of the ERE method in Example 5.3. (a) The position of an entity; (b) the representation of domain values; (c)-(d) violation values marked in the environment.

domain $D_1 = \{1,2,3,4,5,6\}$, row_2 represents $D_2 = \{1,2,3,4\}$, and row_3 represents $D_3 = \{1,2,3,4,5\}$. Figure 5.5(c) shows that if entity a_1 stays at $(3,1)$, meaning $x_1 = 3$, according to constraints $x_1 \neq x_2$ and $x_1 > x_3$, it will violate $x_2 = 3, x_3 = 1, x_3 = 2$, and $x_3 = 3$. Therefore, it will contribute one to the violation values at positions $(3,2), (1,3), (2,3)$, and $(3,3)$. Figure 5.5(d) presents a snapshot for the state of the ERE system with the violation values. Since all entities are at zero-positions, a solution is found.

In Example 5.4, we have divided 4 variables into 2 groups: $\{x_1, x_2\}$ and $\{x_3, x_4\}$. The Cartesian product of each group is $\{\{True, True\}, \{True, False\}, \{False, True\}, \{False, False\}\}$. Figure 5.6(a) shows the domain value of each cell in Example 5.4. If entity a_1 stays at $(1,1)$ (Figure 5.6(b)), it means $x_1 = True$ and $x_2 = True$. According to the constraint set, if entity a_2, for representing variables x_3 and x_4, stays at $(1,2)$, constraint $T(\neg x_1 \vee \neg x_2 \vee \neg x_3) = True$, which is constructed by variables x_1 and x_2 in group $\{x_1, x_2\}$ and variable x_3 in group $\{x_3, x_4\}$, will be violated. If a_2 stays at $(2,2)$, two constraints, $T(\neg x_2 \vee \neg x_3 \vee x_4) = True$ and $T(\neg x_1 \vee \neg x_2 \vee \neg x_3) = True$, will be violated. If a_2 stays at $(3,2)$, it will violate constraint $T(\neg x_1 \vee \neg x_2 \vee x_3) = True$. Furthermore, if a_2 stays at $(4,2)$, constraint $T(\neg x_1 \vee \neg x_2 \vee x_3) = True$ will be violated again. Therefore, the violation

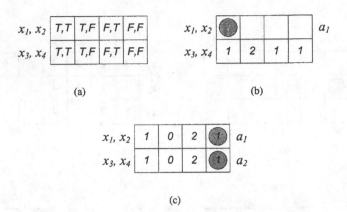

(a)   (b)

(c)

Figure 5.6. An illustration of the ERE method in Example 5.4. (a) The representation of domain values; (b) violation values if entity a_1 is placed on (1, 1); (c) violation values of the whole environment.

values of cells in row_2 are $1, 2, 1$, and 1, respectively. Figure 5.6(c) presents a snapshot for the state of the ERE system with the violation values. Since there are two entities at positions with violation value 1, it is not a solution state.

5.3.3   ERE Entities

All entities inhabit in an environment, in which their positions indicate the value combinations of a variable group. During the operation of the ERE system, entities will keep on moving, based on their primitive behaviors and behavioral rules. At each step, the positions of entities provide an assignment to all variables, whether it is consistent or not. Entities attempt to find better positions that can lead them to a solution state.

Below is a summary of the main policies concerning entity modeling in ERE:

1. $\forall i \in [1, u]$, a_i represents a variable group, $\{x_{i1}, x_{i2}, \cdots, x_{ik}\}$.

2. Entities reside and behave in environment E. Entity a_i can move only to its right or left in row_i, but not up or down. $a_i.j$ represents its x-coordinate, which corresponds to the jth value combination in $D_{i1} \times D_{i2} \times \cdots \times D_{ik}$. Therefore, the position of a_i can be denoted as $(a_i.j, i)$.

In ERE, we introduce function Ψ to define the movement outcome of an entity.

Definition 5.8 (Entity movement function) *Entity movement function Ψ is a function that provides the following mapping:*

$$\Psi : [1, u] \times [1, |D_{i1} \times D_{i2} \times \cdots \times D_{ik}|] \rightarrow [1, |D_{i1} \times D_{i2} \times \cdots \times D_{ik}|]. \quad (5.1)$$

$\Psi(j, i)$ *returns the x-coordinate of a_i at its new position, after it moves from position (j, i). Thus, the new position can be denoted as $(\Psi(j, i), i)$.*

3. In any state of the ERE system, the positions of all entities indicate an assignment to all variables. $\forall i \in [1, u]$, $e(a_i.j, i).value = \langle v_{i1}, v_{i2}, \cdots, v_{ik} \rangle$, that means $x_{i1} = v_{i1}, x_{i2} = v_{i2}, \cdots, x_{ik} = v_{ik}$. By returning the positions of all entities, we can obtain a complete assignment to all variables. If an assignment satisfies all the constraints, i.e., $\forall i \in [1, u]$, $e(a_i.j, i).violation = 0$, it is a solution.

4. a_i can sense the local environment of row_i and perceive the violation value of each cell in row_i. It can find the minimum violation value.

   Here, we define a function $\Phi(i)$ for returning a position (x-coordinate) with the minimum violation value in row_i.

   Definition 5.9 (Minimum position) *A minimum-position is position (j, i) where $i \in [1, u]$, and $\forall j' \in [1, |D_{i1} \times D_{i2} \times \cdots \times D_{ik}|]$, $e(j, i).violation \leq e(j', i).violation$.*

   Definition 5.10 (Minimum position function) *Minimum position function is a function that returns the first minimum-position for entity a_i in row_i:*

   $$\Phi : [1, u] \to [1, |D_{i1} \times D_{i2} \times \cdots \times D_{ik}|], \qquad (5.2)$$

   that is, $\Phi(i) = j$, where (j, i) is a minimum-position, and $\forall j' \in [1, j)$, (j', i) is not a minimum-position.

5. Each entity has its own primitive behaviors. As more than one primitive behavior coexists, each primitive behavior is assigned a probability. Before an entity moves, it will first probabilistically decide the behavior to perform. In other words, the behavioral rule of each entity is probability-based.

6. In what follows, we will describe three primitive behaviors of entities in an ERE system in detail.

   ■ **Least-move**

   An entity moves to a minimum-position with a probability of *least-p*. If there exists more than one minimum-position, the entity chooses the first one on the left of the row. This primitive behavior is instinctive in all entities. A least-move behavior can be expressed as follows:

   $$\Psi(j, i) = \Phi(i). \qquad (5.3)$$

   In this function, the result has nothing to do with the current position j, and the maximum number of computational operations to find the position for each i is $|D_{i1} \times D_{i2} \times \cdots \times D_{ik}|$. We use a special symbol to represent this movement:

$$\Psi_{-l}(j, i) = \Phi(i). \tag{5.4}$$

- **Better-move**
An entity moves to a position, which has a smaller violation value than its current position, with a probability of *better-p*. To do so, it will randomly select a position and compare its violation value with that of its current position, and then decide whether or not to move to this new position. We use function $Random(k)$, which complies with the uniform distribution, to get a random number between 1 and k. A better-move behavior can be defined using function Ψ_{-b}:

$$\Psi_{-b}(j, i) = \left\{ \begin{array}{ll} j, & \text{if } e(r, i).violation \geq e(j, i).violation, \\ r, & \text{if } e(r, i).violation < e(j, i).violation, \end{array} \right. \tag{5.5}$$

where $r = Random(|D_{i1} \times D_{i2} \times \cdots \times D_{ik}|)$.

Although it may not be the best choice for an entity, the computational cost required for this primitive behavior is less than that of least-move. Only two operations are involved in performing this primitive behavior, i.e., producing a random number and performing a comparison. This behavior can easily find a position to move to, if the entity currently stays at a position with a large violation value.

- **Random-move**
An entity moves randomly with a probability of *random-p*. *random-p* will be relatively smaller than the probabilities of selecting least-move and better-move behaviors. It is somewhat like random walk in local search. For the same reason as in local search, random-move is necessary because without randomized movements an ERE system will possibly get stuck in local optima, where all entities are at minimum-positions, but not all of them at zero-positions. In the state of local optima, no entity can move to a new position if using the behaviors of least-move and better-move alone. Thus, the entities will lose their chance for finding a solution if without any techniques to avoid getting stuck in, or to escape from, local optima.

A random-move behavior can be defined using function Ψ_{-r}:

$$\Psi_{-r}(j, i) = Random(|D_{i1} \times D_{i2} \times \cdots \times D_{ik}|). \tag{5.6}$$

The above three primitive behaviors are elementary, simple, and easy to implement. We can combine these primitive behaviors to get complex ones. We will discuss this issue in the later part of this chapter.

By performing a primitive behavior, an entity changes not only its own state (i.e., position), but also the state of its environment (i.e., the violation values). Assume that an entity moves from (j_1, i) to (j_2, i). The violation values of the environment will be updated according to the following two updating rules:

- **Updating-rule 1: Remove-From** (j_1, i):
  For $(\forall i' \in [1, u])(\forall j' \in [1, |D_{i'1} \times D_{i'2} \times \cdots \times D_{i'k}|])$:
  If there are v constraints: (1) they are based on variables included in row_i and $row_{i'}$ and (2) their values are changed from false to true
  Then $e(j', i').violation \leftarrow e(j', i').violation - v$.
- **Updating-rule 2: Add-To** (j_2, i):
  For $(\forall i' \in [1, u])(\forall j' \in [1, |D_{i'1} \times D_{i'2} \times \cdots \times D_{i'k}|])$:
  If there are v constraints: (1) they are based on variables included in row_i and $row_{i'}$ and (2) their values are changed from true to false
  Then $e(j', i').violation \leftarrow e(j', i').violation + v$.

7. In an ERE system, the interactions between entities and their environment are carried out through the above three primitive behaviors. That is, each movement of an entity by performing any primitive behavior will change the violation values stored in the environment. These changes will in turn cause the movements of other entities. In this way, the indirect interactions among entities are implemented through their environment.

8. The goal of each entity in an ERE system is twofold: (1) to find and stay at a zero-position and (2) through indirect interactions, to assist other entities to stay at zero-positions. Each entity, because it can only sense its local environment and cannot sense those of other entities, does not know what a solution is and what the whole system is to achieve. It simply behaves according to its own goal and behavioral rule. That is sufficient for solving a CSP, since if all entities stay at zero-positions, we have found an exact solution to the problem.

5.3.4 System Schedule

The multi-entity system described in this chapter is discrete in nature, with respect to its space, time, and state space. The system will use a discrete timer to synchronize its operations, as illustrated in Figure 5.7.

step = 0:

The system is initialized. We place u entities onto the environment, a_1 in row_1, a_2 in $row_2, \cdots, a_u$ in row_u. The simplest way to place the entities is to randomly select positions. That is, for a_i, we get a position of $(Random($

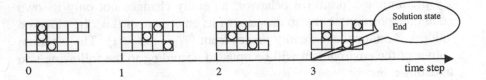

Figure 5.7. Entity movements at different steps.

$|D_{i1} \times D_{i2} \times \cdots \times D_{ik}|), i)$. The initialized positions may correspond to a solution as shown in Figure 5.5(d).

step ← step + 1:

After initialization, the system will start to run.

At each step, which means one unit increment of the system timer, all entities have a chance to decide their movements, i.e., whether or not to move and where to move, and then move synchronously. It should be pointed out that in this chapter, we only concern with a simulation of the multi-entity system, which dispatches entities one by one. The order of dispatching does not influence the performance of the system. It may be based on a random or a predefined sequence.

End:

After the movements of entities, the system will check whether or not all entities are at zero-positions. If so, a solution state is reached. The system will stop and output the solution. Otherwise, the system will continue to dispatch entities to move in the dispatching order.

We can also set a threshold *t-max* for the timer such that when the step reaches *t-max*, the system will stop and output an assignment of the current state, no matter whether or not it is a solution. Another way to terminate the operation is when q entities are staying at zero-positions. Obviously, these settings are for obtaining an approximate solution.

Algorithm 5.3 shows the complete algorithm for an ERE system.

5.3.5   Computational Cost

In what follows, we will examine the space and time complexity of the ERE algorithm in solving CSPs and SATs.

5.3.5.1   Space Complexity

Theorem 5.1 *The space complexity of the ERE algorithm is* $O(\sum_{i=1}^{u} |D_{i1} \times D_{i2} \times \cdots \times D_{ik}|)$.

Algorithm 5.3 The ERE algorithm.

$step = 0$;
Initialize positions and behavior probabilities of entities;
Initialize domain values and violation values of environment;
while *true* **do**
    for all $a_i \in A$ **do**
        Probabilistically determine a primitive behavior, b, to perform;
        Perform primitive behavior b;
        New position $(j'', i) = (\Psi(a_i.j, i), i)$;
        if current position $(a_i.j, i) = (j'', i)$ **then**
            Stay;
        else
            $a_i.j = j''$;
        end if
    end for
    Update violation values of environment;
    if current state satisfies predefined stopping criterion **then**
        Output variable values corresponding to entity positions;
        break;
    end if
    $step + +$;
end while

Proof: The main contribution of this algorithm to space complexity is from domain value storage for all positions in the environment. Assume that we divide all variables into u groups and the ith variable group is $\{x_{i1}, x_{i2}, \cdots, x_{ik}\}$. There will be $|D_{i1} \times D_{i2} \times \cdots \times D_{ik}|$ positions that need to store their domain values. Thus, it needs to store in total $\sum_{i=1}^{u} |D_{i1} \times D_{i2} \times \cdots \times D_{ik}|$ domain values for the whole problem. Therefore, the space complexity is $O(\sum_{i=1}^{u} |D_{i1} \times D_{i2} \times \cdots \times D_{ik}|)$.

The above theorem can further lead to the following specialized theorem for solving SATs.

Theorem 5.2 *In solving an SAT problem, if n variables are equally divided into u groups, the space complexity can be reduced to $O(2^{\lceil \frac{n}{u} \rceil})$.*

Proof: If the variables are divided into u groups. There exist two cases, i.e., $Mod(n, u) = 0$ and $Mod(n, u) \neq 0$.

- $Mod(n, u) = 0$: In this case, the domain value sets corresponding to different variable groups are the same. For example, assume $X = \{x_1, x_2, x_3, x_4\}$, $n = 4$, and we equally divide it into two groups, $\{x_1, x_2\}$ and $\{x_3, x_4\}$.

The corresponding Cartesian products are the same, i.e., $\{\{True, True\},$ $\{True, False\}, \{False, True\}, \{False, False\}\}$. In other words, the domain value sets corresponding to these two variable groups are $\{\{True,$ $True\}, \{True, False\}, \{False, True\}, \{False, False\}\}$. Therefore, it needs to store only one domain value set. Because each domain value set has $2^{\lceil \frac{n}{u} \rceil}$ elements, the space complexity is $O(2^{\lceil \frac{n}{u} \rceil})$.

- $Mod(n, u) \neq 0$: In this case, the first $u - 1$ variable groups have the same number of variables, $\lceil \frac{n}{u} \rceil$. But the uth variable group has fewer variables than the first $u - 1$ groups, i.e., $Mod(n, u)$. For the first $u - 1$ variable groups, they only need to store one domain value set. The uth variable group needs to store its domain value set separately. Therefore, the space complexity is $O(2^{\lceil \frac{n}{u} \rceil} + 2^{Mod(n,u)})$. Since $2^{Mod(n,u)} < 2^{\lceil \frac{n}{u} \rceil}$, the space complexity is still $O(2^{\lceil \frac{n}{u} \rceil})$.

5.3.5.2 Time Complexity

The time complexity of the ERE algorithm is mainly from two steps, i.e., the step to count the number of unsatisfied constraints or clauses and the step to select a position to move to. At these two steps, the time complexity is as follows:

Theorem 5.3 *The time complexity of the step to count the number of unsatisfied constraints is $O(m)$, where m is the number of constraints.*

Proof: To count the number of unsatisfied constraints the ERE algorithm needs to check all m constraints once. Thus, the time complexity is $O(m)$.

Theorem 5.4 *The time complexity of the step to select a position to move to is $O(m \cdot |D_{i1} \times D_{i2} \times \cdots \times D_{ik}|)$.*

Proof: In the ERE method, there are three primitive behaviors for selecting a position to move to, i.e., random-move, better-move, and least-move. Therefore, there are three cases. First, in the case of random-move, the ERE algorithm only needs to randomly select a position, and then move to it without any other processes. Its time complexity is $O(1)$. Secondly, in the case of better-move, the ERE algorithm randomly selects a position, and then evaluates this position. The time complexity to randomly select a position is $O(1)$. To evaluate a position, the ERE algorithm needs to consider the values of all m constraints, that is, its time complexity is $O(m)$. Therefore, the total time complexity is $O(m)$. Thirdly, in the case of least-move, an entity must evaluate all its positions and then select a position with the best evaluation value to move to. Assume that we divide all n variables into u groups and the ith variable group is $\{x_{i1}, x_{i2}, \cdots, x_{ik}\}$, then there will be $|D_{i1} \times D_{i2} \times \cdots \times D_{ik}|$

positions to evaluate. The time complexity to evaluate one position is $O(m)$. Thus, the total time complexity is $O(m \cdot |D_{i1} \times D_{i2} \times \cdots \times D_{ik}|)$. From the above three cases, we can readily conclude that the time complexity to select a position to move to is $O(m \cdot |D_{i1} \times D_{i2} \times \cdots \times D_{ik}|)$.

Corollary 5.1 *In solving an SAT problem, if all variables are equally divided into u group, then the time complexity to select a position is $O(m \cdot 2^{\lceil \frac{n}{u} \rceil})$.*

5.4. An Illustrative Example

In this section, we will walk through an example to show how to apply the ERE method to solve an SAT problem.

Example 5.5 *An SAT is given as follows:*

$$X = \{x_1, x_2, x_3, x_4, x_5\}, n = 5.$$

$$D = \{D_1, D_2, D_3, D_4, D_5\}, D_1 = D_2 = D_3 = D_4 = D_5 = \{True, False\}.$$

$C = \{$

$T(x_3 \vee x_4 \vee \neg x_5)$	$=$	$True,$
$T(x_2 \vee \neg x_3 \vee \neg x_5)$	$=$	$True,$
$T(\neg x_1 \vee \neg x_2 \vee x_3)$	$=$	$True,$
$T(\neg x_1 \vee \neg x_2 \vee x_4)$	$=$	$True,$
$T(\neg x_3 \vee x_4 \vee x_5)$	$=$	$True,$
$T(x_1 \vee \neg x_2 \vee \neg x_3)$	$=$	$True,$
$T(\neg x_2 \vee x_4 \vee x_5)$	$=$	$True,$
$T(\neg x_1 \vee \neg x_3 \vee \neg x_5)$	$=$	$True,$
$T(x_2 \vee \neg x_3 \vee x_4)$	$=$	$True,$
$T(\neg x_1 \vee x_4 \vee \neg x_5)$	$=$	$True,$
$T(x_2 \vee x_3 \vee x_5)$	$=$	$True,$
$T(x_1 \vee x_2 \vee \neg x_4)$	$=$	$True,$
$T(\neg x_1 \vee x_2 \vee \neg x_5)$	$=$	$True,$
$T(\neg x_1 \vee x_3 \vee x_4)$	$=$	$True,$
$T(x_1 \vee \neg x_4 \vee \neg x_5)$	$=$	$True$

$\}$.

In Example 5.5, SAT contains 5 variables and 15 clauses. First, we divide 5 variables into 3 groups: $\{x_1, x_2\}$, $\{x_3, x_4\}$, and $\{x_5\}$, and use 3 entities, $a_1, a_2,$ and a_3, to represent them. Secondly, we model the variable domains as the environment of entities. The domain values will be recorded as $e(j, i).value$ (see Figure 5.8(a)). After that, entities will be randomly placed onto different rows (see Figure 5.8(b)) and the violation values for all positions will be initialized according to the current positions of entities (see Figure 5.8(c)). Thereafter, the cycles of distributed entity movements start. In ERE, entities are dispatched in a random or predefined order. Here, we assume that the order is: $a_1 \to a_2 \to a_3$.

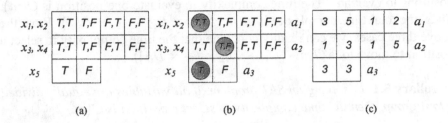

Figure 5.8. An illustration of the ERE method in Example 5.5. (a) Domain values; (b) an initialized state; (c) violation values.

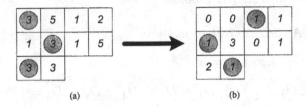

Figure 5.9. The first step in the ERE process. a_1, a_2, and a_3 perform least-move, least-move, and random-move behaviors, respectively.

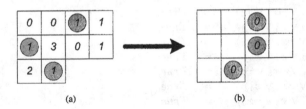

Figure 5.10. The second step in the ERE process. a_1 selects a better-move behavior, but fails to move. a_2 performs a least-move. a_3 also selects a least-move behavior, but fails to move.

At the first step, with respect to the above dispatching sequence, the system first dispatches entity a_1 to move. Entity a_1 moves by performing a least-move behavior, $\Psi_{-l}(1,1) = 3$. As a result, it moves to position $(3,1)$. Entity a_2 takes a least-move too, from $(2,2)$ to $(1,2)$. And, entity a_3 randomly moves from $(1,3)$ to $(2,3)$ (see Figure 5.9(b)). Then, the system checks whether this state is a solution state, and finds that all three entities are not at zero-positions. Therefore, it is not a solution state.

At the second step, entity a_1 selects a better-move, $\Psi_{-l}(3,1) = 4$. But because $(4,1).violation = (3,1).violation = 1$ (see Figure 5.10(a)), a_1 fails to move. Hence, it stays at $(3,1)$. Entity a_2 selects a least-move behavior and moves from $(1,2)$ to $(3,2)$. Entity a_3 also selects a least-move, but fails to move. Hence, it stays (see Figure 5.10(b)). Next, the system finds that all entities are at zero-positions, which means it is at a solution state:

a_1 stays at position $(3, 1) \Rightarrow \{x_1 = False, x_2 = True\}$;

a_2 stays at position $(3, 2) \Rightarrow \{x_3 = False, x_4 = True\}$;

a_3 stays at position $(2, 3) \Rightarrow \{x_5 = False\}$.

Hence, the final solution is: $x_1 = False, x_2 = True, x_3 = False, x_4 = True$, and $x_5 = False$.

5.5.   Experimentation

In the preceding sections, we have described the ERE method. In this section, we will present several ERE experiments on n-queen problems and benchmark SAT problems. We will also discuss some important issues related to ERE.

In the experiments, we will initialize all entities with the same parameters *random-p*, *least-p*, and *better-p* ($\forall i \in [1, u]$, $a_i.random\text{-}p = random\text{-}p$, $a_i.least\text{-}p = least\text{-}p$, $a_i.better\text{-}p = better\text{-}p$).

5.5.1   N-Queen Problems

Let us first examine the performance of the ERE method in solving n-queen problems.

Experiment 5.1 *This experiment examines how well the ERE system performs in the first three steps.* $n=\{500, 1000, 1500, 2000, 2500, 3000, 3500, 4000, 4500, 5000, 5500, 6000, 6500, 7000\}$. *The size of a variable group is 1. least-p:random-p = n. type = $F2BLR$[1] (10 runs)(see Table 5.1).*

Observation 5.1 *From Table 5.1, we note that:*

1. *After initialization, nearly 10% of the entities stay at zero-positions (although this is not shown in Table 5.1).*

2. *After the 1st step, nearly 80% of the entities stay at zero-positions. That means 80% of the assignments to variables can satisfy constraints. This result is very significant because it is obtained with just one step.*

3. *After the 2nd step, nearly $n - 23$ entities stay at zero-positions. That means about $n - 23$ assignments to variables can satisfy constraints. In other words, only about 23 assignments cannot satisfy constraints.*

4. *After the 3rd step, nearly $n - 7$ entities stay at zero-positions. This is a good approximate solution obtained in just three steps, no matter how large n is.*

---

[1]It will be discussed in Section 5.6.1.

Table 5.1. Performance in the first three steps of ERE-based n-queen problem solving.

Step	1	2	3
n	Number of entities at zero-positions	Ratio of entities at zero-positions	Number of entities not at zero-positions
500	388.9	0.78	17.9	4.8
1,000	796.6	0.80	23.4	8.6
1,500	1,206.7	0.80	22.2	7.9
2,000	1,601.9	0.80	22.2	7.8
2,500	1,999.9	0.80	22.2	3.9
3,000	2,398.7	0.80	24.4	7.0
3,500	2,808.1	0.80	23.0	6.3
4,000	3,204.4	0.80	26.0	6.5
4,500	3,589.9	0.80	24.1	7.7
5,000	3,990.4	0.80	25.5	8.3
5,500	4,417.0	0.80	22.7	7.5
6,000	4,775.5	0.80	23.5	8.4
6,500	5,221.2	0.80	21.5	8.1
7,000	5,567.1	0.80	20.9	7.6
Average		0.8	22.8	7.2

5.5.2   Benchmark SAT Problems

In order to examine the performance of ERE in solving SATs, we will run ERE on some benchmark problems from [SATLIB, 2000]: uf100-430 and flat50-115. Hoos and Stützle made an empirical evaluation of different local search algorithms for SAT [Hoos and Stützle, 2000a], and two of their testsets are uf100-430 and flat50-115. Uf100-430 is a subset of Uniform Random-3-SAT problems [SATLIB, 2000] where each clause of instances has exactly three literals randomly selected from $2n$ literals according to a uniform distribution. (Here, we assume that there are n variables to construct instances.) Flat50-115 is a subset of Flat Graph Coloring problems [SATLIB, 2000] where clauses may have a different number of literals.

The ERE algorithm for SATs is slightly different from the one for n-queen problems. In SAT, we equally divide variables into several groups. In the following experiments, we tune the optimal parameter settings first, and then collect statistical results on uf100-430 and flat50-115.

Experiment 5.2 *This experiment aims at examining the effectiveness and efficiency of the ERE method in solving benchmark SAT problems with exact solutions. The testset is a subset of Uniform Random-3-SAT problems: uf100-430. It includes 1000 instances, and each instance contains 100 variables*

and 430 clauses. The size of a variable group is 4. least-p:random-p = 40.
type = F2BLR (100 runs).

Experiment 5.3 *This experiment addresses the efficiency of the ERE method in solving benchmark SAT problems with exact solutions. The testset is a subset of Flat Graph Coloring problems: flat50-115. It includes 1000 instances, and each instance contains 150 variables and 545 clauses. The size of a variable group is 3. least-p:random-p = 80. type = F2BLR (100 runs).*

Table 5.2. Mean-movement(flip)-numbers in benchmark SATs.

Algorithms	uf100-430	flat50-115
GWSAT	6,532	7,023
GSAT/TABU	4,783	1,040
HWSAT	3,039	2,641
WalkSAT	3,672	3,913
WalkSAT/TABU	2,485	61,393
Novelty	28,257	20,065
R-Novelty	1,245	7,109
ERE	2,950	3,548

Observation 5.2 *In the last row of Table 5.2, we have given our experimental results, i.e., the mean number of movements for entities to get solutions in 100 runs and for 1000 different instances. Also in Table 5.2, we have shown some experimental data as given in [Hoos and Stützle, 2000a]. We note that:*

1. *As compared to other popular algorithms in the SAT community, our ERE method presents comparable results on both testsets.*

2. *For some algorithms in Table 5.2, their performances are not stable in different testsets. For example, the performance of R-Novelty is the best in testset uf100-430, but in testset flat50-115. In this respect, our ERE method yields better results, i.e., it is stable in two different problem types.*

5.5.2.1   Fair Measurement

One important issue that we should clarify is the rationale behind the comparison between the movements in the ERE algorithm and the flips in other algorithms. In the algorithms listed in Table 5.2, the number of flips is an important and commonly used index to evaluate the performance. 'Flip' means changing the value of a variable in a complete assignment from $True$ to $False$, or from $False$ to $True$. In the ERE method, a movement of an entity will cause, in an extreme case, n/u variables to change their values. If we just consider the value changes that occur in the variables, the comparison in Table 5.2 is unfair to the other algorithms. In essence, what those algorithms actually count is how many steps they take to get a solution rather than the values changed (in the other algorithms, the number of steps is equal to the number of flips, because they are sequential). In this sense, the correct way to compare the ERE method and the other algorithms is to compare their steps. But, because the ERE algorithm as implemented in this chapter is a sequential simulation of the actual ERE algorithm, we recorded the movements of all entities in Table 5.2 and compared them with the flips in the other algorithms. Although one movement may cause multiple flips, all these flips happen simultaneously. Thus, it is fair in this case to compare movements with flips.

5.5.2.2   Performance Evaluation

In what follows, we will examine the performance of the ERE method in finding 'approximate' solutions to SATs. For an SAT problem, there are three possible answers: 'satisfiable', 'unsatisfiable', and 'unknown'. Here, we employ the notion of an 'approximate' solution to include the case in which clauses are partially satisfied under a complete assignment to all variables. Through the following experiments, we will see that ERE is efficient in finding an 'approximate' solution in the first three steps.

Experiment 5.4 *This experiment examines how well the ERE method performs in the first three steps to find approximate solutions. Testsets are five subsets of Uniform Random-3-SAT problems:* $\{uf50, uf100, uf150, uf200, uf250\}$. *The testsets include 1000, 1000, 100, 100, and 100 instances, respectively. The number of variables ranges from 50 to 250, and the corresponding number of clauses from 218 to 1065. We give each instance 10 runs, and at the same time, we calculate the mean value for the number of satisfied clauses at each step in the first three steps (see Table 5.3).*

Experiment 5.5 *This experiment examines how efficient the ERE method is in finding approximate solutions. Testsets are four subsets of Flat Graph Coloring problems:* $\{flat50, flat100, flat200, flat250\}$. *The first testset includes 1000 instances. The last three testsets include 100 instances. The number of variables ranges from 150 to 600, and the number of clauses from 545 to 2237. We give each instance 10 runs, and calculate the mean value for the number of satisfied clauses at each step in the first three steps (see Table 5.4).*

In Experiments 5.4 and 5.5, all instances are satisfiable. Let us see a special situation, that is, all instances are unsatisfiable.

Experiment 5.6 *This experiment examines how efficient the ERE method is in finding approximate solutions. In this experiment, as in Experiment 5.4, test-sets are from Uniform Random-3-SAT problems:* {$uuf50$, $uuf100$, $uuf150$, $uuf200$, $uuf250$}. *The other parameters are the same as Experiment 5.4 except that all instances in this experiment are unsatisfiable (see Table 5.5).*

Table 5.3. The number of satisfied clauses and its ratio to the total number of clauses in Experiment 5.4. Note: S_C is Satisfied Clauses. Ratio is the ratio between S_C and the total number of clauses. The same is true for Tables 5.4 and 5.5.

Step	1		2		3	
Testset	S_C	Ratio	S_C	Ratio	S_C	Ratio
uf50	208	0.954	212	0.972	213	0.977
uf100	410	0.953	418	0.972	420	0.977
uf150	615	0.953	628	0.974	630	0.977
uf200	820	0.953	836	0.972	840	0.976
uf250	1,016	0.954	1,036	0.972	1,040	0.977

Table 5.4. The number of satisfied clauses and its ratio to the total number of clauses in Experiment 5.5.

Step	1		2		3	
Testset	S_C	Ratio	S_C	Ratio	S_C	Ratio
flat50	513	0.941	534	0.980	536	0.983
flat100	1,051	0.941	1,093	0.980	1,097	0.982
flat150	1,581	0.941	1,644	0.979	1,650	0.982
flat200	2,103	0.940	2,189	0.979	2,197	0.982

Observation 5.3 *From Tables 5.3, 5.4, and 5.5, we can observe that with the ERE method:*

1. *We can get an approximate solution with about 94-95% satisfied clauses after the 1st step, no matter whether or not the instance is satisfiable. This is a better result than that of ERE in solving n-queen problems where about 80% entities are at zero-positions.*

2. *After the 2nd step, the number of satisfied clauses has increased rapidly. But, the amount of increase is different in the two different types of testset. The increase in Flat Graph Coloring is greater.*

Table 5.5. The number of satisfied clauses and its ratio to the total number of clauses in Experiment 5.6.

Step	1		2		3	
Testset	S_C	Ratio	S_C	Ratio	S_C	Ratio
uuf50	208	0.954	211	0.968	212	0.972
uuf100	409	0.951	417	0.970	419	0.974
uuf150	614	0.952	626	0.970	629	0.975
uuf200	819	0.952	835	0.971	838	0.974
uuf250	1,014	0.952	1,034	0.971	1,038	0.975

3. *After the 3rd step, the ratio of satisfied clauses will reach 97-98% no matter whether or not the instance is satisfiable.*

Based on the above observation, we can say that ERE is efficient and robust.

5.6.   Discussions

In this section, we will comment on some design and engineering issues as related to the ERE method. Also, we will highlight the distinct features of the ERE method, as compared to other CSP solving methods.

5.6.1   Necessity of the Better-Move Behavior

Generally speaking, in previous search algorithms, there are two types of flip: greedy flip (i.e., flipping will cause the steepest hill-climbing) and random flip. But, in our ERE method, there are three primitive behaviors: least-move, better-move, and random-move. That is to say, ERE has a new primitive behavior, i.e., better-move. Is it necessary?

Although better-move and least-move may seem to be similar (i.e., to move to a position based on the violation value), they are essentially different. At each step, it would be much easier for an entity to use a least-move to find a better position to move to than to use a better-move. This is because a least-move checks all the positions within a row, whereas a better-move randomly selects and checks only one position. If all entities use only random-move and better-move behaviors, the efficiency of the system will be low, because many entities cannot find a better position to move to at each step. On the other hand, the time complexity of a better-move is much less than that of a least-move. Therefore, if we can find a good balancing point between a better-move and a least-move, we can greatly improve the performance.

In ERE, we have found a way to balance these two behaviors. First, an entity will use a better-move to compute its new position. If it succeeds, the entity will move to the new position. If it fails, it will continue to perform some further better-moves until it finds a successful better-move. If it fails to take all better-moves, it will then perform a least-move. But, the question that remains is how many better-moves before a least-move will be desirable.

It is obvious that at the initialization step, many entities are not at 'good' positions, i.e., they are at positions with large violation values. In this case, the probability of using a better-move to successfully find a position to move to is high. But, as the process goes on, more and more entities will be at good positions. At this time, there will be less chance for an entity to move using a better-move behavior.

In order to confirm the above proposition, we have further conducted two experiments. Our experimental results show that the ERE algorithm will yield the best performance if there are two better-moves before a least-move. More better-moves will increase the runtime. On the other hand, fewer better-moves cannot provide enough chance for entities to find better positions. Also, we find that a $F2BLR$ behavior can obtain the best performance. Here, $F2BLR$ means that at the first step, an entity will probabilistically select a least-move or random-move behavior to perform. If it chooses a least-move behavior to perform, it will have two chances to select a better-move before performing a least-move. However, this is the case for the first step only.

5.6.2   Probability Setting

Among three primitive behaviors in the ERE method, the least-move and better-move behaviors play important roles in the performance of ERE. However, the random-move is still necessary, because if without random-move, i.e., *random-p*=0, the system may get stuck in local optima.

Now, here is a question: how to set the probabilities for the three primitive behaviors in order to have the best performance of ERE? From Section 5.6.1, we can see that the better-move behavior occurs as a prologue of the least-move in the $F2BLR$ behavior. In this case, the combination of better-move and least-move will have the same probability as a single least-move. Therefore, the most important probabilities are *least-p* and *random-p*. It is the ratio of *least-p* to *random-p* that plays an important role in the system. Our experimental results show that when the ratio is $1.5n$ for n-queen problems or about $1.5u$ for SAT problems, the performance of the ERE algorithm will be the best.

5.6.3   Variable Grouping

In the ERE method, we divide variables into groups. In the experiments given in Sections 5.5.1 and 5.5.2, we have divided variable into groups of an equal size, i.e., one variable forms a group in an n-queen problem, and four or five variables are grouped together in an SAT problem. Through the experiments, we note that how to partition variables is a very important factor in the performance of the ERE algorithm. We have observed from some experiments that which variables should be placed into a group is a more important aspect than the size of a variable group. However, to date, how to partition the variables in an optimal way still remains unsolved.

5.6.4   Characteristics of ERE

In Sections 5.5.1 and 5.5.2, we have shown that the ERE method is efficient in finding an exact solution to n-queen problems, and more stable than other SAT algorithms. Also in these sections, we have seen that the performance of ERE in finding an approximate solution is very efficient. After the first three steps, about $n - 7$ queens are at zero-positions in n-queen problems, and about 97-98% clauses are satisfied in SAT problems. This property is quite important if a solution is required with a hard deadline. Now, let us take a look at some interesting characteristics of ERE:

1. An ERE system is a self-organizing system. The process of solving CSPs by ERE is entirely determined by the locality and parallelism of individual entities. Thus, the computation required is low.

2. The movement of an entity may affect the whole environment. And, the change in the environment will in turn affect the movements of other entities. In other words, the interactions among entities are indirectly carried out through the medium of their environment. In this sense, we may regard that the entities cooperate with each other in finding a solution.

3. The ERE method is quite open and flexible. We can easily add new primitive behaviors or combine present primitive behaviors into complex ones. In addition, we may give each entity different parameter settings, as well as modify the entity-environment interaction according to the specific requirements of the problems.

5.6.5   Comparisons with Existing Methods

In this subsection, we will compare the ERE method to the existing heuristic or distributed methods, and highlight their distinct features.

5.6.5.1   Comparison with Min-Conflicts Heuristics

The ERE method differs from the min-conflicts method in the following aspects:

1. In the min-conflicts hill-climbing system reported in [Minton et al., 1992], the system chooses a variable at each step that is currently in conflict and reassign its value by searching the space of possible assignments and selecting the one with the minimum total conflicts. The hill-climbing system can

get trapped in a local minimum (note that the same phenomenon can also be observed from the GDS network for constraint satisfaction). On the other hand, in our method, an entity is given a chance to select a random-move behavior according to its probability, and hence it is capable of escaping from a local trap. In our present work, we also note that the extent to which the entities can most effectively avoid the local minima and improve their search efficiency is determined by the probabilities (i.e., behavior selection probabilities) of the least-move and random-move behaviors.

2. Another system introduced in [Minton et al., 1992] is informed backtracking. It arguments a standard backtracking method with the min-conflicts ordering of the variables and values. This system attempts to find a sequence of repairs, such that no variable is repaired more than once. If there is no way to repair a variable without violating a previously repaired variable, the algorithm backtracks. It incrementally extends a consistent partial assignment in the same way as a constructive backtracking program, however, it uses information from the initial assignment to guide its search. The key distinction between this method and ours is that our method does not require backtracking. As stated by Minton et al. [Minton et al., 1992], their system trades search efficiency for completeness; for large-scale problems, terminating in a no-solution report will take a very long time.

3. In both min-conflicts hill-climbing and informed backtracking systems proposed in [Minton et al., 1992], the key is to compute and order the choice of variables and values to consider. It requires to test all related constraints for each variable and to test all its possible values. This step is similar to the Remove-From and Add-To operations in our method, except that we only test a selected position and do not sort the variables. The use of the ordering heuristic can lead to excessive assignment evaluation preprocessing and therefore will increase the computational cost at each step.

4. In our present method, we examine the use of a fewer-conflicts repair, by introducing the better-move behavior, that requires only one violation value evaluation for each variable. The empirical evidence has shown that the use of the high-priority better-move when combined with other behaviors can achieve more efficient results. We believe that the reason that using the currently-available min-conflicts value at each step can compromise the systems performance is because the min-conflicts values quickly reduce the number of inconsistencies for some variables but at the same time also increase the difficulties (e.g., local minima) for other variables.

5.6.5.2   Comparison with Distributed Constraint Satisfaction

Our multi-entity method has several fundamental distinctions from Yokoo et al.'s distributed constraint satisfaction method, as listed below:

1. Yokoo et al.'s method does not require a global broadcasting mechanism or structure. It allows agents to communicate their constraints to others by sending and receiving messages such as *ok?*, and *nogood*. In other words, their methods handle the violation checking among agents (variables) through agent-to-agent message exchanges, such that each agent knows all instantiated variables relevant to its own variables.

   In our method, the notion of entity-to-entity communication is implicit – we assume that for violation updating, each entity (representing the value of a variable or a group of variables) is 'informed' about the values from relevant entities (representing the values of relevant variables) either by means of accessing an $n \times n$ look-up memory table or via pairwise value exchange – both implementations enable an entity to obtain the same information, but the latter can introduce significant communication overhead costs (i.e., longer cycles required [Yokoo et al., 1998]) to the entities.

2. In the asynchronous weak-commitment search algorithm [Yokoo et al., 1998, Yokoo and Hirayama, 2000], a consistent partial solution is incrementally extended until a complete solution is found. When there exists no value for a variable that satisfies all the constraints between the variables included in the partial solution, this algorithm abandons the whole partial solution and then constructs a new one. Although asynchronous weak-commitment search is more efficient than asynchronous backtracking, abandoning partial solutions after one failure can still be costly. In the case of the ERE method, the high-level control mechanism for maintaining or abandoning consistent partial solutions is not required.

   Yokoo et al. have also developed a non-backtracking algorithm called distributed breakout, which provides a distributed implementation for the conventional breakout [Yokoo and Hirayama, 2000].

3. In asynchronous weak-commitment search, each agent utilizes the min-conflicts heuristic as mentioned in Section 5.6.5.1 to select a value from those consistent with the agent_view (those values that satisfy the constraints with variables of high-priority agents, i.e., value-message senders).

   On the other hand, the ERE method utilizes a combination of value-selection heuristics that involves a better-move behavior for efficiently finding fewer-conflicts repairs.

4. As related to the above two remarks, the asynchronous weak-commitment search and asynchronous backtracking algorithms are designed to achieve

completeness and thus the steps of backtracking and incremental solution constructing or abandoning are necessary, whereas the ERE method is aimed at more efficiently finding an approximate solution, which is useful when the amount of time available for an exact solution is limited.

5. Last but not the least, we have also systematically compared the performance of the ERE system with that of Yokoo et al.'s algorithms, namely, asynchronous backtracking, asynchronous backtracking with min-conflicts heuristic, and asynchronous weak commitment, in solving some benchmark n-queen problems [Yokoo et al., 1998]. We can establish that as demonstrated in solving the benchmark n-queen problems, ERE is an effective method and the number of cycles used in the ERE system is very competitive with those by Yokoo et al.'s method, given that our formulation utilizes different behavior prioritization and violation checking schemes.

In summary, as complementary to each other, both Yokoo et al.'s asynchronous method and the ERE method can be very efficient and robust when applied in the right context. A distinct feature of Yokoo et al.'s asynchronous method is, like other standard backtracking techniques, its completeness, whereas the feature of the ERE method lies in its efficiency and robustness in obtaining an approximate solution within a few steps (although it empirically always produces an exact solution when enough steps are allowed). The ERE method is not guaranteed to be complete since it involves random-moves. Another feature of the ERE method is that its behaviors are quite easy to implement.

5.6.5.3 Remarks on Partial Constraint Satisfaction

Partial constraint satisfaction is a very desirable way of solving CSPs that are either over constrained or too difficult to solve [Wallace, 1996]. It is also extremely useful in situations where we want to find the best solution obtainable within fixed resource bounds or in real-time. Freuder and Wallace are the pioneers in systematically studying the effectiveness of a set of partial constraint satisfaction techniques using random problems of varying structural parameters [Freuder and Wallace, 1992]. The investigated techniques included basic branch and bound, backjumping, backmarking, pruning with arc consistency counts, and forward checking. Based on the measures of constraint checks and total time to obtain an optimal partial solution, forward checking was found to be the most effective. Also of general interest is that their work has offered a model of partial constraint satisfaction problems (PCSPs) involving a standard CSP, a partially ordered space of alternative problems, and a notion of distances between these problems and the original CSP.

Our present work attempts to develop, and empirically examine, an efficient technique that is capable of generating partial constraint satisfaction solutions.

This work shares the same motivation as that of Freuder and Wallace's work [Freuder and Wallace, 1992, Wallace, 1996], and also emphasizes that the costs of calculating (communicating) and accessing constraint violation information should be carefully considered in developing a practically efficient technique. That is also part of the reason why much attention in our work has been paid to (1) the use of environmentally updated and recorded violation values (without testing from the scratch for each variable) and (2) the effectiveness of the better-move behavior in finding an approximate solution.

5.6.5.4   Remarks on Entity Information and Communication for Conflict-Check

Yokoo et al.'s method and the ERE method have a common thread; both formulations employ multiple entities that reside in an environment of variables and constraints (although in the ERE method, the environment also contains violation information, which is analogous to the 'artificial pheromone' in an ant system [Dorigo et al., 1999, Dorigo et al., 1991]) and make their own decisions in terms of how the values of certain local variables should be searched and selected in the process of obtaining a globally consistent solution.

Nevertheless, it should be pointed out that the present implementations of the two methods differ from each other in the way in which the entities record and access their conflict-check information. The former utilizes a sophisticated communication protocol to enable the entities representing different groups of variables and constraints to exchange their values. By doing so, the entities are capable of evaluating constraint conflict status with respect to other relevant entities (variables). On the other hand, our implementation utilizes a feature of entity current-value broadcast to enable other entities to compare with their values and to update the violation values in their local environment. Although the formulations may seem different, the objectives as well as effects of them are fundamentally similar. The reasons that we decided to use value broadcast and sharing are threefold: First, the implementation can make use of a common storage space and by doing so avoid introducing the same space requirement to every entity. Secondly, it can reduce the overhead costs incurred during the pairwise information exchange, which can be quite significant. Thirdly, since our ERE method extensively uses fewer-conflicts moves, such behaviors can be triggered based on only one possible violation evaluation instead of n assignment evaluations, and hence the access to such a broadcast information source is not demanding.

5.6.5.5   Remarks on Sequential Implementation

Theoretically, the ERE method can be implemented in a parallel fashion. However, for the sake of illustration, we have used a sequential implementation to simulate the multi-entity concurrent or synchronous actions and to test the effectiveness of our method. Our sequential simulation utilizes a global simulated clock, called step. The state of the environment as well as the entities (i.e., the positions of the entities) will be changed only at each discrete step. In order to simulate the concurrent or synchronous actions of the entities at step k, we let the individual entities perform their cycles of behavior selection, value selection, and violation updating. In so doing, the entities are dispatched in a sequential fashion. Once this is completed, the state of the system will then be refreshed with the new positions of the entities corresponding to the newly-selected values, and thereafter, the simulated clock will be incremented to $k + 1$. Here, it is worth mentioning that apart from the fact that our implementation simulates the operations of a parallel system, the empirical results of our sequential ERE implementation are still comparable to those reported in [Minton et al., 1992, Yokoo et al., 1998], if we evaluate the performance using the measures of number of constraint checks, as introduced by Freuder and Wallace [Freuder and Wallace, 1992], and space complexity.

5.7. Entity Network for Complexity Analysis

As we have described, ERE is a multi-entity based method for distributed problem solving. Generally speaking, in multi-entity systems for problem solving, entities will implicitly or explicitly form an entity network that connects all interacting entities. As an entity may or may not cooperate, coordinate, or compete with other entities, the resulting entity network may not be fully connected. Hence, two straightforward questions can be raised: What is the topology of a network formed by entities? How does the resulting entity network reflect the computational complexity of a given problem? Answering these questions will help us understand the performance of the multi-entity system. It will also guide us in designing more reasonable methods for solving problems. In this section, we will address the above questions in the context of our ERE method.

First, let us define the concept of an entity network.

Definition 5.11 (Entity network) *An entity network is a virtual[2] graph corresponding to a multi-entity system, where vertices are entities, and edges (also called links) are the implicit or explicit relationships of cooperation, coordination, or competition among entities.*

Correspondingly, the topology of an entity network can be defined as follows:

---

[2]The word 'virtual' implies that there may not exist physical links among entities in a multi-entity system.

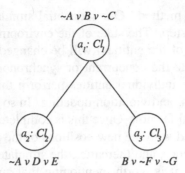

Figure 5.11. Representing three clauses into an entity network, where each vertex denotes an entity, and each edge denotes a common variable shared by two corresponding entities. $a_i{:}Cl_i$ denotes that entity a_i represents clause Cl_i.

Definition 5.12 (Entity network topology) *The topology of an entity network is the geometrical property of the network, which reflects the connectivity of the vertices and edges.*

5.7.1   Different Representations

When using a multi-entity system to solve CSPs, there are two different types of representation: constraint-based representation, where entities represent constraints, and variable-based representation, where entities represent variables. As we have seen from the preceding sections, ERE utilizes a variable-based representation. What we are interested in here is the topology of an entity network under different representations. Specifically, in what follows, we will examine which representation is better based on some benchmark SAT problems.

5.7.1.1   Clause Based Representation

In a clause-based representation, an entity represents a clause in a given SAT problem. In order to satisfy a clause, an entity needs to assign values to the variables in this clause such that at least one literal is true. Because a variable can appear in multiple clauses simultaneously, the entities that have common variables should cooperate and find consistent values to the variables. In this case, if two entities have a common variable, we regard it as an edge between the corresponding entity vertices.

Figure 5.11 presents a schematic diagram of a clause-based entity network. In this figure, entities a_1 and a_2 have a common variable A, and entities a_1 and a_3 have a common variable B. Hence, there are two edges between entities a_1 and a_2 and between entities a_1 and a_3, respectively. Because entities a_2 and a_3 have no identical variable, there is no edge between them.

Experimental Testing. Based on the above representation scheme, we have conducted some experiments using benchmark SAT problems: Uniform-3-SAT and Flat Graph Coloring problems [SATLIB, 2000]. We randomly selected 10% of the instances from each testset. As two important measures for characterizing the topology of an entity network, the average characteristic path length, L_G, and clustering coefficient, C_G, of the network were calculated for each testset, as shown in Table 5.6.

Here, characteristic path length L_G and clustering coefficient C_G of entity network G are defined as follows [Watts and Strogatz, 1998]:

Given an entity network, $G = \langle V, R \rangle$, where $V = \{v_1, v_2, \cdots, v_n\}$ is a set of entities and $R = \{r_1, r_2, \cdots, r_m\}$ is a set of edges between entities in V.

1. Characteristic path length:

$$L_G = \frac{2}{n \cdot (n-1)} \sum_{i,j \in \{1,\cdots,n\}, i \neq j} d_{i,j}, \qquad (5.7)$$

where $d_{i,j}$ is the shortest distance between entities a_i and a_j.

2. Clustering coefficient:

$$C_G = \frac{1}{n} \sum_{i=1}^{n} c_{a_i}, \qquad (5.8)$$

where c_{a_i} is the clustering ratio (also called clustering in short) of entity a_i. Assume $d(a_i)$ is the degree of a_i (i.e., the number of neighboring entities of a_i), and $b(a_i)$ is the number of existing edges between the neighbors of a_i. Therefore,

$$c_{a_i} = \frac{b(a_i)}{\frac{(d(a_i)+1) \cdot d(a_i)}{2}} = \frac{2 \cdot b(a_i)}{(d(a_i)+1) \cdot d(a_i)}. \qquad (5.9)$$

In general, L_G is a global property of entity network G that indicates the connectivity of G. C_G, on the other hand, is a local property that reflects the average connectivity of cliques in G. In essence, C_G denotes the possibility that two entities, which have a common neighboring entity, are neighbors.

Small World Topology. Milgram first proposed the notion of small world [Milgram, 1967]. Later, Watts and Strogatz mathematically formulated a small world topology based on the means of characteristic path length and clustering coefficient [Watts and Strogatz, 1998]. Small world phenomena have been extensively found in natural systems (e.g., human society [Milgram, 1967], food Web in ecology [Montoya and Sole, 2000]) as well as in man-made systems (e.g., the World Wide Web [Adamic, 1999]). Walsh observed such phenomena

Table 5.6. Clause-based representation for benchmark SATs: Uniform-3-SAT and Flat Graph Coloring. In the experiments, we randomly selected 10% of the instances from each testset, and calculated the average values of L_G, C_G, and μ for the selected instances. For each selected instance, we generated 10 random entity networks with the same number of vertices and edges. We calculated the average L_G (i.e., L_{random}) and C_G (i.e., C_{random}) of these random entity networks in order to compare them with those of the corresponding instances.

Testset	Vertices	Edges	L_G	L_{random}	C_G	C_{random}	μ
Uf50	218	4,072	1.836	1.829	0.444	0.215	2.071
Uf75	325	6,125	1.923	1.895	0.417	0.162	2.614
Uf100	430	8,113	2.007	1.945	0.407	0.135	3.106
Uf125	538	10,240	2.077	1.992	0.399	0.118	3.514
Uf150	645	12,276	2.146	2.040	0.393	0.107	3.854
Uf175	753	14,366	2.205	2.086	0.389	0.099	4.148
Uf200	860	16,434	2.260	2.130	0.387	0.093	4.408
Uf225	960	18,194	2.309	2.175	0.384	0.089	4.587
Uf250	1,065	20,168	2.352	2.215	0.383	0.085	4.768
Flat30	300	1,883	3.196	2.533	0.576	0.183	3.962
Flat50	545	3,826	3.433	2.670	0.567	0.155	4.694
Flat75	840	6,092	3.673	2.791	0.564	0.144	5.154
Flat100	1,117	8,072	3.875	2.893	0.564	0.141	5.368
Flat125	1,403	10,258	4.019	2.965	0.565	0.137	5.583
Flat150	1,680	12,160	4.155	3.045	0.563	0.137	5.619

in search problems, such as graph coloring, time tabling, and quasigroup problems [Walsh, 1999]. He further experimentally showed that the small world topology could make a search process very difficult.

Watts and Strogatz defined that a graph G with n vertices and m edges has a small world topology if and only if:

$$L_G \approx L_{random} \quad \text{and} \quad C_G \gg C_{random}, \tag{5.10}$$

where L_{random} and C_{random} are the average characteristic path length and clustering coefficient of random graphs with the same size as G (i.e., n vertices and m edges). Here, G must be connected, i.e., $k \gg ln(n)$, where $k = \frac{2m}{n}$ is the average degree of vertices in G [Watts and Strogatz, 1998].

The above definition of a small world is qualitative. In order to measure the "small worldiness" of a graph, Walsh provided a quantitative measurement, i.e., proximity ratio, μ [Walsh, 1999]:

$$\mu = \frac{\frac{C_G}{L_G}}{\frac{C_{random}}{L_{random}}} = \frac{C_G \cdot L_{random}}{C_{random} \cdot L_G}. \tag{5.11}$$

A small world topology requires $\mu \gg 1$. The larger the μ, the more "small worldy" the graph (i.e., the graph has more clusters).

Clause: $\sim A \vee B \vee \sim C$

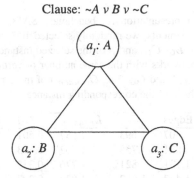

Figure 5.12. Representing a clause into an entity network, where each entity corresponds to one variable. $a_i{:}X$ denotes that entity a_i represents variable X.

To test whether or not there exist small world topologies in the entity networks obtained from the clause-based representation of SATs, we calculated the average L_{random} and C_{random} as well as μ for each testset (see Table 5.6). Corresponding to each instance in a testset, we generated 10 random entity networks with the same size and calculated their C_{random} and L_{random}. The results show that in all testsets, $L_G \approx L_{random}$, $C_G \gg C_{random}$, and $\mu > 2$. This means that with the clause-based representation, the resulting entity networks have small world topologies.

5.7.1.2 Variable Based Representation

In SATs, each clause normally contains several literals. In order to satisfy a clause, the related variables should be assigned compatible values in order to guarantee that at least one literal is true. In this sense, clauses act as constraints among variables. Since ERE represents variable groups with entities, the constraints among variables are implicitly transformed into constraints among entities. In order to satisfy a constraint (i.e., a clause), entities that represent the variables in the constraint will restrain each other. If we assume each entity represents only one variable, the entity network can be constructed as follows: A vertex denotes an entity. An edge exists between two entities if and only if the corresponding variables appear in a certain clause simultaneously.

Figure 5.12 provides a schematic diagram of a variable-based entity network, where variables A, B, and C are in the same clause, $\neg A \vee B \vee \neg C$. There is an edge between each pair of entities.

Experimental Testing. Based on the above representation scheme, we encoded some benchmark SAT problems, Uniform-3-SAT and Flat Graph Coloring problems [SATLIB, 2000], into entity networks. By calculating their characteristic path lengths and clustering coefficients, we examined whether

Table 5.7. Variable-based representation for benchmark SATs: Uniform-3-SAT and Flat Graph Coloring. In these experiments, we randomly selected 10% of the instances from each testset, and calculated average L_G, C_G, and μ of the selected instances. For each instance, we generated 10 random entity networks with the same number of vertices and edges. We calculated average L_G (i.e., L_{random}) and C_G (i.e., C_{random}) of these random entity networks in order to compare them with those of the corresponding instances.

Testset	Vertices	Edges	L_G	L_{random}	C_G	C_{random}	μ
Uf50	50	507	1.693	1.587	0.501	0.469	1.140
Uf75	75	831	1.758	1.702	0.403	0.360	1.154
Uf100	100	1,131	1.821	1.776	0.335	0.294	1.168
Uf125	125	1,457	1.862	1.822	0.297	0.255	1.190
Uf150	150	1,780	1.900	1.860	0.271	0.227	1.219
Uf175	175	2,095	1.939	1.895	0.250	0.207	1.238
Uf200	200	2,414	1.968	1.925	0.237	0.191	1.267
Uf225	225	2,727	2.001	1.957	0.223	0.179	1.276
Uf250	250	3,037	2.031	1.987	0.214	0.169	1.297
Flat30	90	270	3.662	2.677	0.387	0.334	1.585
Flat50	150	495	3.935	2.836	0.350	0.296	1.641
Flat75	225	765	4.186	3.025	0.337	0.279	1.671
Flat100	300	1,017	4.375	3.179	0.330	0.274	1.659
Flat125	375	1,278	4.532	3.290	0.332	0.269	1.697
Flat150	450	1,530	4.647	3.390	0.327	0.268	1.676
Flat175	525	1,776	4.735	3.479	0.329	0.267	1.674
Flat200	600	2,037	4.849	3.539	0.325	0.265	1.681

or not there existed small world topologies. Table 5.7 presents our experimental findings.

Non-Small-World Topology.   From Table 5.7, we note that for all testsets of Uniform-3-SAT, $L_G \approx L_{random}$, $C_G \approx C_{random}$, and $1.1 < \mu < 1.3$. This indicates that small world topologies do not exist. For all testsets of Flat Graph Coloring, $L_G \not\approx L_{random}$, $C_G \not\approx C_{random}$, and μ is around 1.65. They do not show small world topologies, either. Thus, we can conjecture that with a variable-based representation, the resulting network does not have a small world topology.

In the above encoding scheme, an entity can also represent several variables. In this case, two entities will be connected by an edge if and only if two variables, respectively represented by these two entities, appear in an identical clause. The resulting entity network can be derived from the entity network where each entity represents one variable by merging some of the original entities into a single one. Obviously, the former network is denser than the latter one. A small world topology normally exists in a connected, but 'sparse', network. Since the latter entity network does not have a small world topology, the former will not either. This point has been demonstrated in some other experiments, where μ is still less than 2.0.

5.7.2 Discussions on Complexity

The previous sections have experimentally examined the topologies of entity networks as obtained from two different representations of some given problems. In this section, we will further discuss their computational complexities. By doing so, we will understand the rationale behind the variable-based representation in the ERE method. We will then obtain a guiding principle for designing a multi-entity system for problems solving.

5.7.2.1 Complexities under Different Representations

From the previous section, we note that given an SAT problem, different representations can lead to different entity networks. In the two cases mentioned, the first one shows small world topologies in its resulting entity networks. The second one, i.e., ERE, does not generate small world topologies.

Walsh has studied the relationship between topologies (in particular, small world topologies) and computational complexities. In [Walsh, 1999], Walsh empirically verified that a small world topology increases the computational complexity of a search algorithm that involves certain heuristics. This is because heuristics normally guide a search process locally. But in a small world network, based on local information an algorithm cannot well predict global properties of the problem at hand.

We have also experimentally validated this finding based on the previous two representations of SAT problems. Specifically, using the same SAT problems, we have examined a clause-based representation as opposed to a variable-based representation as used in ERE. In our experiments, each entity represents a clause. A clause acts as the local space where the corresponding entity resides and behaves. A literal in a clause is a position where an entity can stay. If an entity stays at a certain position, the corresponding literal will be true. If two entities stay at two positions whose literals are negations of each other, a conflict occurs. For the whole system, to solve a given SAT is to eliminate conflicts. Our experimental results suggest that with such a clause-based representation, it is normally hard (in term of entity movements) to solve a problem, which is however relatively easier to solve by ERE with a variable-based representation.

5.7.2.2 Balanced Complexities in Intra- and Inter-Entity Computations

In ERE, an entity can represent one or more variables. Section 5.7.1.2 has experimentally shown that in both cases, the resulting entity networks do not have small world topologies. But, which case is better? Experiments have suggested that as the number of variables represented by an entity in ERE increases, the resulting entity networks will be less 'small worldy', because the networks become smaller and denser. In an extreme case, we may use only one entity to represent all variables. In this case, the network becomes to an isolated vertex. Is this the best situation?

The answer is no. In fact, if an entity represents multiple variables, the entity should assign values to all its variables. Because the variables of an entity are usually not independent, as the number of variables represented by the entity increases, the intra-entity computational complexity to assign values to its variables increases, too. Therefore, a good design should balance the intra- and inter-entity computational complexities to achieve the lowest total computational cost.

With respect to this point, our experiments have suggested that when an entity represents four or five variables, the total computational cost, in terms of entity movements and variable flips, becomes the lowest.

5.7.2.3   A Guiding Principle

Based on the above discussions, we can arrive at a guiding principle for designing a multi-entity based method for solving a problem:

- It should avoid having a small world topology in its resulting entity network.

- It should maintain balanced intra- and inter-entity computational complexities in order to achieve the lowest total computational cost.

5.8.   Summary

In this chapter, we have described an AOC-by-fabrication based, multi-entity method, called ERE, for solving constraint satisfaction problems (CSPs), such as n-queen problems, and satisfiability problems (SATs), which are regarded as special cases of CSPs in this chapter. Specifically, we have presented the general model of the ERE method. Through experiments on benchmark problems, we have empirically examined the performance of ERE. Next, we have discussed several important issues in ERE, such as probability setting and variable grouping. Finally, by studying the topologies of entity networks as formed in ERE, we have shown the rationale behind the variable-based representation in ERE and have obtained a guiding principle for designing multi-entity based problem solving systems. The following are some specific remarks on ERE and AOC-by-fabrication.

5.8.1   Remarks on ERE

The key ideas behind ERE rest on three notions: Environment, Reactive rules, and Entities. In ERE, each entity can only sense its local environment and apply some behavioral rules for governing its movements. The environment records and updates the local values that are computed and affected according to the movements of individual entities. In solving a CSP with the ERE method, we first divide variables into several groups, and then represent each variable group with an entity whose positions correspond to the elements of the Cartesian product of variable domains. The environment for the whole multi-entity system contains all the possible domain values for the problem, and at the same time, it also records the violation values for all the positions.

An entity can move within its row, which represents its domain. So far, we have introduced three primitive behaviors: least-move, better-move, and random-move. Using these primitive behaviors, we can constitute other complex behaviors, such as $F2BLR$. The movement of an entity will affect the violation values in the environment. It may add or reduce the violation value of a position. After being randomly initialized, the ERE system will keep on dispatching entities, according to a certain predefined order, to choose their movements until an exact or approximate solution is found.

Besides describing the ERE method, we have also experimentally demonstrated how this method can effectively find a solution and how efficient it is in solving various n-queen problems for both exact and approximate solutions. Several practical rules for parameter settings have been discussed and established following our observations from the experiments. Furthermore, we have employed ERE to solve SATs and have obtained some experimental results on benchmark SAT testsets [Hoos and Stützle, 2000b, Liu and Han, 2001]: Uniform Random-3-SAT and Flat Graph Coloring from the SATLIB. Our experimental results are comparable to, and more stable than, those of the existing algorithms.

5.8.2 Remarks on AOC by Fabrication

As we can see from this chapter, the AOC-by-fabrication approach focuses on building a mapping between a real problem and a natural phenomenon or system, the working mechanism behind which is more or less known (see Figure 5.13). In the mapping, the synthetic entities and their parameters (e.g., states and behaviors) correspond respectively to the natural life-forms and their properties. Ideally, some special states of the natural phenomenon or system correspond to the solutions of the real problem.

From the ERE example, we can note the following common characteristics of the AOC-by-fabrication approach:

1. There is a group of autonomous entities, each of which is mainly characterized by sets of goals, states, behaviors, and behavioral rules. Entities may be homogeneous or heterogeneous. Even in the homogeneous case, entities may differ in some detailed parameters.

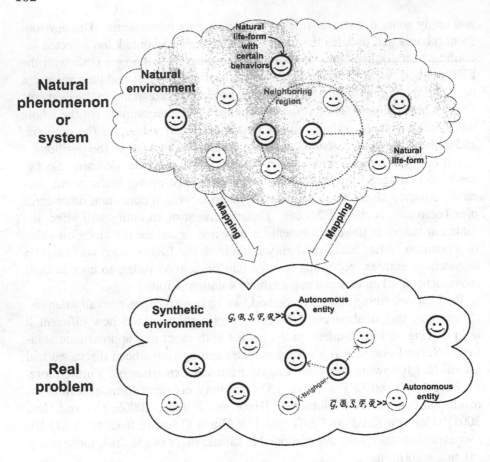

Figure 5.13. The AOC-by-fabrication approach, which is intended to build a mapping between a real problem and a natural phenomenon or system. In the figure, entities are characterized by a group of parameters (e.g., $\mathcal{G}$, $\mathcal{B}$, $\mathcal{S}$, $\mathcal{F}$, and $\mathcal{R}$), the meanings of which have been given in Chapter 4 (Definitions 4.3-4.9).

2. The composition of the entity group may change over time, through the process analogous to birth (amplification of the desired behavior) and death (elimination of the undesired behavior). But, in some applications, the number of entities is fixed.

3. The interactions between autonomous entities are local; neither global information nor central executive control is needed.

4. The environment is dynamical and records the information related to the current status of the problem. It serves as a medium for information sharing among autonomous entities.

5. The local goals of autonomous entities drive the selection of their primitive behaviors at each step.

6. The global goal of the whole AOC system is represented by a universal fitness function that measures the progress of the computation.

Exercises

5.1 Give brief answers to the following questions:

   (a) What are the key features of the ERE method?

   (b) Why can the ERE method successfully and efficiently solve some computationally hard problems?

5.2 ERE is an AOC-based method for solving constraint satisfaction problems as well as satisfiability problems. According to the AOC framework discussed in Chapter 4, formulate the ERE method.

5.3 What are the similarities and differences between ERE and local search?

5.4 A graph coloring problem can be described as follows: Given a graph $G = \langle V, E \rangle$, where V is a node set and E is an edge set, you are required to use k colors to paint all nodes such that no two adjacent nodes have the same color. Graph coloring problems fall into a category of classical constraint satisfaction problems. Try to use ERE to solve graph coloring problems.

5.5 In the described ERE method,

   (a) Probabilities *least-p*, *better-p*, and *random-p* are empirically determined. Find an adaptive or self-adaptive way to determine the probabilities?

   (b) Variable grouping is predefined: Variables are orderly divided into groups with almost the same size. This way of grouping does not consider the inherent relationships among the variables belonging to the same group. Find other ways to group variables in order to improve the efficiency of the ERE method. The following are two hints:

   - Relation-based grouping: Study relationships among variables and group variables based on them;
   - Dynamical grouping: Variable groups are dynamically changed.

5.6 Tang et al. introduced an entity compromise technique in [Tang et al., 2003], where neighboring entities compromise and act as a single entity in order to eliminate constraints among them. Try to apply this technique to the ERE method.

5.7 Think about other techniques (e.g., self-adaptation) to improve the ERE method.

Chapter 6

AOC in Complex Systems Modeling

6.1. Introduction

Although the Internet, in particular, the World Wide Web, brings deeper and deeper influence on people's daily life, the regularities and working mechanisms behind many Internet related phenomena still remain unknown. Therefore, how to reveal and characterize such regularities and mechanisms becomes a pressing mission to computer scientists.

Complex systems modeling is another goal of AOC. In this chapter, we will show how the AOC-by-prototyping approach is used to solve a complex systems modeling task, i.e., revealing the unknown factors that determine the regularities in Web surfing [Liu et al., 2004b]. In so doing, we can note the features of the AOC-by-prototyping approach.

In the real world, with the help of a blueprint, engineers can build a model of a system in an orderly fashion. When there is insufficient knowledge about the mechanism showing how the system works, it is difficult, if not impossible, to build such a model. Assumptions about the unknown workings have to be made in order to get the process started. Given some observable behavior of the desired system, designers can verify the model by comparing that of the model with the desired features. The process will be repeated several times before a good, probably not perfect, prototype is found. This is AOC-by-prototyping. Apart from obtaining a working model of the desired system, an important byproduct of the process is the discovery of the mechanisms that are unknown when the design process first started. This view is shared by researchers developing and testing theories about society and social phenomena [Conte and Gilbert, 1995, Doran and Gilbert, 1994]. Other studies have used similar methods, for instance, in the problem domains of highway traffic flow [Helbing and Huberman, 1998], traffic jams [Rasmussen and Barrett, 1995, Howard, 1997], crowd control in sports grounds [Still, 2000], and crowd dynamics in a smoky room where a fire has broken out [Helbing et al., 2000a].

6.1.1   Regularity Characterization

Researchers have recently observed several interesting, self-organized regularities from the World Wide Web, ranging from the structure and growth of the Web to the access patterns in Web surfing. What remains to be a great challenge in Web log mining is how to explain user behavior underlying observed Web usage regularities. By experimenting with the entity-based decision model of Web surfing, we aim to explain how some Web design factors as well as user cognitive factors may affect the overall behavioral patterns in Web usage.

Viewing the Web as a large directed graph of nodes (i.e., Web pages) connected with links (i.e., hyperlinks), Huberman et al. proposed a random-walk model to simulate certain regularities in user navigation behavior and suggested that the probability distribution of surfing depth (steps) follows a two-parameter inverse Gaussian distribution [Huberman et al., 1997]. They conjectured that the probability of finding a group surfing at a given level scales inversely in proportion to its depth, i.e., $P(L) \sim L^{-3/2}$, where L is depth.

In order to further characterize user navigation regularities as well as to understand the effects of user interests, motivation, and content organization on user behavior, in this chapter we will present an information foraging entity based model that takes into account the interest profiles, motivation aggregation, and navigation strategies of users.

6.1.2   Objectives

The random-walk model [Huberman et al., 1997, Lukose and Huberman, 1998] and the Markov chain model [Levene et al., 2001, Levene and Loizou, 1999] have been used to simulate statistical regularities as empirically observed from the Web. However, these models do not relate the emergent regularities to the dynamical interactions between users and the Web, nor do they reflect the inter-relationships between user behavior and the contents or structure of the Web. They are, by and large, black-box methods that do not explicitly address the details of interacting entities.

The issues of user interest and motivation to navigate on the Web are among the most important factors that directly determine user navigation behavior [Thatcher, 1999]. In our present study, we aim to take one step further by proposing a new AOC model of Web surfing that takes into account the characteristics of users, such as interest profiles, motivations, and navigation strategies. By doing so, we attempt to answer the following questions:

1. Is it possible to experimentally observe regularities similar to empirical Web regularities if we formulate the aggregation of user motivation? In

other words, is it possible to account for empirical regularities from the point of view of motivation aggregation?

2. Are there any navigation strategies or decision making processes involved that determine the emergence of Web regularities, such as the distributions of user navigation depth?

3. If the above is validated, will different navigation strategies or decision making processes lead to different emergent regularities? In other words, when we observe different power-law distributions, can we tell what are dominant underlying navigation strategies or decision making processes that have been used by users?

4. What is the distribution of user interest profiles underlying emergent regularities?

5. Will the distribution of Web contents as well as page structure affect emergent regularities?

6. If we separately record users who can successfully find relevant information and those who fail to do so, will we observe different regularities?

In order to answer the above questions, we will develop a white-box, AOC model. This model should, first of all, incorporate the behavioral character-istics of Web users with measurable and adjustable attributes. Secondly, it should exhibit the empirical regularities as found in Web log data. Thirdly, the operations in the model should correspond to those in the real-world Web surfing.

In the next section, we will present our white-box, information foraging entity based model for characterizing emergent Web regularities. Foraging entities are information seeking entities that are motivated to find certain infor-mation of their special interest from the pages in artificial Web space.

6.2. Background

This section provides an overview of research work related to Web mining. Generally speaking, Web mining is aimed to study the issues of (1) where and how information can be efficiently found on the Web and (2) how and why users behave in various situations when dynamically accessing and using the information on the Web.

6.2.1 Web Mining for Pattern Oriented Adaptation

The first major task in Web mining may be called Web mining for pattern oriented adaptation, that is to identify the inter-relationships among different websites, either based on the analysis of the contents in Web pages or based on the discovery of the access patterns from Web log files. By understanding such inter-relationships, we aim to develop adaptive Web search tools that help facilitate or personalize Web surfing operations.

This task is certainly justified as studies have shown that 85% of users use search engines to locate information [GVU, 2001]. Even though good search engines normally index only about 16% of the entire Web [Lawrence and Giles, 1999], an adaptive utility can still be very useful to filter or rank thousands of Web pages that are often returned by search engines. For instance, some researchers have developed efficient search techniques that detect authorities, i.e., pages that offer the best resource of the information on a certain topic, and hubs, i.e., pages that are collections of links to authorities [Cbakrabarti et al., 1999, Gibson et al., 1998]. When it is difficult to directly find relevant information from search engines, navigating from one page to another by following a hyperlink has become a natural way of searching for information. In this respect, it will be even more important to adaptively organize Web information in such a way that relevant information can be conveniently accessed.

6.2.1.1   Web Data Mining

As classified by Mobasher, Web data mining has traditionally been dealing with three problems: computing association rules, detecting sequential patterns, and discovering classification rules and data clusters [Mobasher et al., 1996]. This classification of Web mining work has its counterparts in the field of data mining. Pitkow summarized the previous work in Web mining with respect to different data sources, such as client, proxy, gateways, server, and Web [Pitkow, 1998]. Cooley presented a taxonomy of Web mining that distinguishes Web content mining from Web usage mining [Cooley et al., 1997].

6.2.1.2   User Behavior Studies

Web usage mining deals with the analysis of Web usage patterns, such as user access statistical properties [Catledge and Pitkow, 1995, Cuhna et al., 1995], association rules and sequential patterns in user sessions [Cooley et al., 1999, Pei et al., 2000, Spiliopoulou, 1999, Spiliopoulou et al., 1999, Zaane et al., 1998], user classification and Web page clusters based on user behavior [Joshi and Krishnapuram, 2000, Nasraoui et al., 1999, Yan et al., 1996]. The results of Web usage mining can be used to understand user habits in browsing information as well as to improve the accessibility of websites.

6.2.1.3   Adaptation

The primary objective of Web mining for pattern oriented adaptation is to help users efficiently surf and retrieve information from the Web. One way to make information search efficient is to reduce the latency in information search by means of optimizing cache algorithms based on user browsing behavior characteristics on proxies or gateways [Barford et al., 1999, Breslau et al., 1998, Glassman, 1994], or by means of prefetching Web contents. Padmanabhan proposed a method of predictive prefetching based on the analysis of user navigation patterns [Padmanabhan and Mogul, 1996]. However, this method is only useful if the relevant information contents at the next level can be correctly predicted [Cuhna and Jaccoud, 1997]. Some studies have examined the issue of website workload [Arlitt and Williamson, 1996, Barford and Crovella, 1998] and network traffic [Mogul, 1995] in order to find ways to improve the efficiency of information response and propagation.

Other examples of Web mining for pattern oriented adaptation include the studies on finding efficient search or personalization algorithms that directly work with the contents on the Web as well as the structure of the Web [Madria et al., 1999, Spiliopoulou, 1999].

6.2.2 Web Mining for Model Based Explanation

The second important task in Web mining can be referred to as Web mining for model-based explanation, that is to characterize user navigation behavior during Web surfing operations, based on empirical regularities as observed from Web log data.

6.2.2.1 Web Regularities

Many identified interesting regularities are best represented by characteristic distributions following either a Zipf-like law [Zipf, 1949] or a power law. That is, the probability P of a variant taking value k is proportional to $k^{-\alpha}$, where α is from 0 to 2. A distribution presents a heavy tail if its upper tail declines like a power law [Crovella and Taqqu, 1999].

What follows lists some of the empirical regularities that have been found on the Web:

1. The popularity of requested and transferred pages across servers and proxy caches follows a Zipf-like distribution [Barford et al., 1999, Breslau et al., 1998, Cuhna et al., 1995, Glassman, 1994].

2. The popularity of websites or requests to servers, ranging from Web user groups to fixed user communities, such as within a proxy or a server, follows a power law [Adamic and Huberman, 2000, Breslau et al., 1998, Maurer and Huberman, 2000].

3. The request inter-arrivals and Web latencies follow a heavy-tail distribution [Barford and Crovella, 1998, Helbing et al., 2000b, Yan et al., 1996].

4. The distribution of document sizes either across the Web or limited to pages requested in a proxy or a certain user community exhibits a heavy tail [Arlitt

and Williamson, 1996, Barford et al., 1999, Barford and Crovella, 1998, Cuhna et al., 1995].

5. The number of pages either across all websites or within a certain domain of the Web follows a power law [Huberman and Adamic, 1999a].

6. The trace length of users within a proxy or a website, or across the Web follows a power law [Adar and Huberman, 2000, Huberman et al., 1997, Levene et al., 2001, Lukose and Huberman, 1998].

7. The dynamical response of the Web to a Dirac-like perturbation follows a power law [Johansen and Sornette, 2000].

8. The distribution of links (both incoming and outgoing) among websites or pages follows a power law [Adamic and Huberman, 1999, Albert et al., 1999, Barabasi and Albert, 1999, Barabasi et al., 2000, Broder et al., 2000].

6.2.2.2   Regularity Characterization

Although researchers have empirically observed strong regularities on the Web, few of them have dealt with the issue of how such regularities are emerged. Some black-box models of regularities consider only the input and output data correspondence for a "system", without explicitly addressing the rationale of underlying mechanisms. In [Barabasi and Albert, 1999], a random network model with growth and preferential attachment factors is proposed that produces a power distribution of link number over websites or pages. Huberman showed that the power-law distribution of page number over various websites can be characterized based on a stochastic multiplicative growth model coupled by the fact that websites appear at different times and grow at different rates [Huberman and Adamic, 1999b]. He also presented a random-walk model to simulate user navigation behavior that leads to a power distribution of user navigation steps [Huberman et al., 1997, Lukose and Huberman, 1998]. Levene developed an absorbing Markov chain model to simulate the power-law distribution of user navigation depth on the Web [Levene et al., 2001, Levene and Loizou, 1999].

6.3.   Autonomy Oriented Regularity Characterization

In our work, we are interested in finding the inter-relationship between the statistical observations on Web navigation regularities and the foraging behavior patterns of individual entities. In what follows, we will introduce the notions and formulations necessary for the modeling and characterization of Web regularities with information foraging entities.

6.3.1 Artificial Web Space

In the entity-based Web regularity characterization, we view users as information foraging entities inhabiting in a particular environment, namely, the Web space. In this section, we will address the characterization issue of the Web space.

The Web space is a collection of websites connected by hyperlinks. Each website contains certain information contents, and each hyperlink between two websites signifies certain content similarity between them. The contents contained in a website can be characterized using a multi-dimensional content vector where each component corresponds to the relative information weight on a certain topic. In order to build the artificial Web space that characterizes the topologies as well as connectivities of the real-world Web, we introduce the notion of an artificial website that may cover contents related to several topics and each topic may include a certain number of Web pages. Such a website may also be linked to other websites of similar or different topics through URLs.

6.3.1.1 Web Space and Content Vector Representations

We consider the Web space as a graph consisting of nodes and links, as suggested in [Broder et al., 2000]. The nodes correspond to websites or pages, whereas the links correspond to hyperlinks between them. The information contents in a certain node are represented using the weights of a content vector as follows:

$$C_n = [cw_n^1, cw_n^2, \ldots, cw_n^i, \ldots, cw_n^M], \tag{6.1}$$

where

C_n: content vector for node n (i.e., website or page);
cw_n^i: relative content information weight on topic i;
M: number of topics.

To determine the content similarity between two nodes, we will make use of the following distance function:

$$d(C_i, C_j) = \left(\sum_{k=1}^{M} (cw_i^k - cw_j^k)^2 \right)^{1/2}, \tag{6.2}$$

where $d(C_i, C_j)$ denotes the Euclidean distance between the content vectors of nodes i and j.

Thus, based on the preceding definition, we are able to specify the relationship between the contents of two nodes. For instance, when two nodes are linked through a hyperlink, it is reasonable to assume that the contents contained in the two nodes is somewhat related, that is to say, their content vector distance is below a certain positive threshold.

6.3.1.2   Modeling Content Distributions

Now that we have defined a means of representing node contents, our next question is how to describe the distribution of node contents with respect to various topics. In our present study, we will investigate the behavior of information foraging entities interacting with Web pages. The contents of those Web pages are distributed following a certain statistical law. Specifically, we will implement and contrast two models of content distribution: normal distribution and power-law distribution.

1. **Normal distribution:** The content weight cw_n^i with respect to topic j in node n is initialized as follows:

$$cw_n^i = \begin{cases} \text{T}+ \mid \text{X}_c \mid, & \text{if } i = j, \\ \mid \text{X}_c \mid, & \text{otherwise,} \end{cases} \tag{6.3}$$

$$f_{\text{X}_c} \sim normal(0, \sigma_p), \tag{6.4}$$

$$\text{T} \sim normal(\mu_t, \sigma_t), \tag{6.5}$$

where
$$\begin{array}{ll} f_{\text{X}_c}: & \text{probability distribution of weight } \text{X}_c; \\ normal(0, \sigma_p): & \text{normal distribution with mean 0 and variance } \sigma_p; \\ \text{T}: & \text{content (increment) offset on a topic;} \\ \mu_t: & \text{mean of normally distributed offset T;} \\ \sigma_t: & \text{variance of normally distributed offset T.} \end{array}$$

In the above model, we assume that all content weights on a topic are non-negative. We can adjust σ_t and μ_t to get various topic distributions in Web pages; the smaller σ_t is or the larger μ_t is, the more focused the node will be on the topic.

2. **Power-law distribution:** In this model, the content weight of node n on topic j, cw_n^i, will follow a power law:

$$cw_n^i = \begin{cases} \text{T}+ \mid \text{X}_c \mid, & \text{if } i = j, \\ \mid \text{X}_c \mid, & \text{otherwise,} \end{cases} \tag{6.6}$$

$$f_{\text{X}_c} \sim \alpha_p(\text{X}_c + 1)^{-(\alpha_p+1)}, \ \text{X}_c > 0, \ \alpha_p > 0, \tag{6.7}$$

where
$$\begin{array}{ll} f_{\text{X}_c}: & \text{probability distribution of weight } \text{X}_c; \\ \alpha_p: & \text{shape parameter of a power-law distribution;} \\ \text{T}: & \text{content (increment) offset on a topic.} \end{array}$$

Similar to the model of a normal distribution, here we can adjust α_p to generate different forms of a power-law distribution.

6.3.1.3 Constructing an Artificial Web

Having introduced the notions of content vector representation and content distribution models, in what follows we will discuss how to add links to the artificial Web space.

There are two major steps involved. First, we create several groups of nodes, where each group focuses on a certain topic. The distribution of the contents in the nodes follows a specific model as given above. We assume that an information entity starts its foraging from a Web homepage that contains links to the nodes of several topics. In our study, we assign this homepage an equal distance to individual topics as follows:

$$cw_p^i = T_c, \quad i = 1, \ldots, M, \tag{6.8}$$

where cw_p^i denotes the content weight of the homepage on topic i. T_c denotes the content (increment) offset on the topic.

After initializing the content vectors, the next step is to build links between the nodes. As mentioned above, we assume that when there is a link between two nodes, the information contents of the nodes should be related. Therefore, we will build a link between two nodes only if the content vector distance between them is below a positive distance threshold, r. r can be adjusted in order to generate Web clusters of different degrees of connectivity. In this respect, we refer to r as the degree-of-coupling of websites. Increasing r leads to increasing the number of links in a website (that is, the similarity between the contents of two linked nodes will decrease).

Algorithm 6.1 summarizes the key steps in constructing the artificial Web space.

Algorithm 6.1 Constructing the artificial Web space.

---
for each topic k **do**
    Create a node group and content vectors;
end for
for each node i in the group **do**
    Initialize the link list of node i;
    for each node j ($j \neq i$) in the group **do**
        if $d(C_i, C_j) < r$ **then**
            Add node j to the link list of node i;
            Add $d(C_i, C_j)$ to the link list of node i;
        end if
    end for
end for

6.3.1.4   Remarks on the Artificial Web Server

In the construction of our artificial Web server, we have assumed that two pages are similar if they are linked. This assumption has been found to be generally valid with respect to the real-world Web by several researchers [Menczer, 2004a, Menczer, 2004b, Flake et al., 2002]. For instance, in the studies reported in [Menczer, 2004b], Menczer has examined the relationship between content, linkage, and semantic similarity measures across a large number of real-world Web page pairs and has found that the Pearson's correlation coefficients between content and linkage similarity measures significantly positive. For instance, the content similarity measure can reach up to $0.4 \sim 0.6$ when the linkage similarity measure (a neighborhood function) is around 0.6. Both measures will have peaks around 0.9. Such a correlation is found to be significantly positive in the Web pages that deal with News, Home, Science, Sports, Reference, and Games among others. In [Menczer, 2004a], Menczer has further formalized and quantitatively validated two conjectures that are often taken for granted; they are:

1. The link-content conjecture that "a page is similar to the pages that link to it," and

2. The link-cluster conjecture that "pages about the same topic are clustered together."

Having said so, it should be pointed out that given the variety of kinds of links that are created in the real-world Websites, "distance" may not always be a good indication of "relevance" among Web pages. In some cases, two Web pages may be linked simply because one adds a special feature or service to another.

6.3.1.5   Dynamically Generated Web Pages

In the real-world Web, some portion of pages may be "hidden" in databases; they are generated on the fly. In the artificial Web pages constructed in this study, we have not considered the dynamical generation of Web pages, but used only existing and continuing pages. Although our virtual Web pages may, to a certain extent, model the characteristics of the dynamically generated Web pages, there are still differences between them that deserve further experimental examinations taking both facets into consideration.

6.3.2   Foraging Entities

In our AOC model, we will use foraging entities to emulate Web users. Therefore, how to characterize foraging entities will be a crucial task in our modeling. In the following, we will address users' interest profile representation, interest distribution, and motivational support aggregation.

6.3.2.1 Interest Profiles

Each entity forages in the Web space with different interests in mind, e.g., accessing a specific website for an update on some contents, searching for information related to some topics, or simply wandering in the Web space to browse various topics. The interest profile of an entity will determine its behavior in Web surfing. In this section, we will describe how to model the interest profile of an entity using a multi-dimensional preference vector that specifies the interests of the entity in various topics. In addition, we will introduce the measure of interest entropy to characterize whether or not an entity has a balanced interest profile.

Specifically, we define the preference vector of an entity as follows:

$$P_m = [pw_m^1, pw_m^2, \ldots, pw_m^i, \ldots, pw_m^M], \quad (6.9)$$

$$p_{mi} = \frac{pw_m^i}{\sum_{j=1}^{M} pw_m^j}, \quad (6.10)$$

$$H_m = -\sum_{i=1}^{M} p_{mi} log(p_{mi}), \quad (6.11)$$

where

P_m: preference vector of entity m;
pw_m^i: preference weight of entity m on topic i;
H_m: interest entropy of entity m.

In Equation 6.11, we define H_m in a similar way as we define the measure of entropy in information theory. Here, H_m indicates the breadth and balance of an entity's interests in different topics. The larger H_m is, the more evenly distributed the entity's interests will be. As a result, the entity is more likely to have multiple goals and jump from one topic to another in its surfing. When the entity has equal interests in all topics, the value of H_m will be the largest, i.e.,

$$H_{max} = -\sum_{i=1}^{M} \frac{1}{M} log \left(\frac{1}{M} \right) = log(M). \quad (6.12)$$

As will be seen in the next section, the quantity of interest entropy will affect the decision of an entity on which Web page it will select among several others.

6.3.2.2 Modeling Interest Distributions

In order to investigate how different interest distributions may influence the behavior patterns in entity foraging, in our study we will specifically implement and observe two interest distribution models: normal distribution and power-law distribution. Thus, the preference vector of a foraging entity will be initialized as follows:

1. **Normal distribution:** The weight of a preference vector, pw_m^i, for entity m on topic i is defined as follows:

$$pw_m^i = X_p, \qquad (6.13)$$

$$f_{X_p} \sim normal(0, \sigma_u), \qquad (6.14)$$

where $normal(0, \sigma_u)$ denotes the normal distribution with mean 0 and variance σ_u.

2. **Power-law distribution:** The probability distribution of entity m's preference weight on topic i, pw_m^i, is given as follows:

$$pw_m^i = X_p, \qquad (6.15)$$

$$f_{X_p} \sim \alpha_u(X_p + 1)^{-\alpha_u+1}, \; X_p > 0, \; \alpha_u > 0, \qquad (6.16)$$

where α_u denotes the shape parameter of a power-law distribution.

We can get various interest profiles of foraging entities by adjusting parameters σ_u and α_u.

6.3.2.3 Motivational Support Aggregation

When an information foraging entity finds certain websites in which the content is close to its interested topic(s), it will become more ready to search the websites at the next level, that is, it will get more motivated to surf deeper. On the other hand, when an entity does not find any interesting information after some foraging steps or it has found sufficient contents satisfying its interests, it will stop foraging and leave the Web space. In order to model such a motivation-driven foraging behavior, here we introduce a support function, S_t, which serves as the driving force for an entity to forage further. When an entity has found some useful information, it will get rewarded, and thus the support value will be increased. As the support value exceeds a certain threshold, which implies that the entity has obtained a sufficient amount of useful information, the entity will stop foraging. In other words, the entity is satisfied with what it has found. On the contrary, if the support value is too low, the entity will lose its motivation to forage further and thus leave the Web space.

Specifically, the support function is defined as follows:

$$S_{t+1} = S_t + \theta \cdot \Delta M_t + \phi \cdot \Delta R_t, \qquad (6.17)$$

where

S_t: support value at step t;
ΔM_t: motivational loss at step t;
ΔR_t: reward received at step t;
θ, ϕ: coefficients of motivation and reward terms, respectively.

The initial support value, maximum and minimum support thresholds will be set, respectively, as follows:

$$rclinit\_support_m = \frac{1}{2}\sum_{i=1}^{M} pw_m^i, \qquad (6.18)$$

$$max\_support_m = \sum_{i=1}^{M} pw_m^i, \qquad (6.19)$$

$$min\_support_m = 0, \qquad (6.20)$$

where pw_m^i denotes the preference weight of entity m with respect to topic i.

6.3.3   Foraging in Artificial Web Space

Generally speaking, hyperlinks from a Web page are connected to other pages covering the same or similar topics. The anchor texts that are associated with the hyperlinks usually indicate the topics of the linked pages. In the process of information foraging, an entity will examine the anchor texts and then predict which of the linked next-level pages may contain more interesting contents. In so doing, the success of content prediction by the entity will depend on its specific navigation strategy.

Earlier research on closed hypertext systems, databases, and library information systems have suggested that there possibly exist three browsing strategies: search browsing (directed search where the objective is known), general purpose browsing (consulting sources that have a high likelihood of items of interest), and serendipitous browsing (purely random) [Cove and Walsh, 1988]. In this section, we will provide the computational models of three navigation strategies that serve as the behavioral rules of information foraging entities. As we will see, these behavioral rules are based on the evaluation of the pages at the next level. According to their navigation strategies, information foraging entities can be classified into three types: random entities, rational entities, and recurrent entities. Moreover, we will describe the primitive behavior of entities for updating their interest profiles, motivation, and reward functions during Web navigation. Finally, we will elaborate on the foraging behavior through an algorithm.

6.3.3.1   Navigation Strategies

Suppose that entity m is currently on page n that belongs to topic j (also referred to as domain here). There are h hyperlinks inside page n, among which

h_1 hyperlinks belong to the same topic as page n and h_2 hyperlinks belong to other topics. We can describe the strategies of foraging entities in selecting the next-level Web page, i.e., selecting a hyperlink, k, from h hyperlinks, in terms of selection probabilities, as follows:

1. **Random entities:** Random entities have no strong interests in any specific topics. They wander from one page to another. In so doing, their decision on selecting the next-level page is random. The probability, p_k, of reaching node k at the next step can be written as follows:

$$p_k = \frac{1}{h}, \quad k = 1, \dots, h. \tag{6.21}$$

2. **Rational entities:** Most foraging entities behave rationally. Rational entities have certain interested topics in mind and they forage in order to locate the pages that contain information on those topics. When they reach a new website, they will try to decide whether or not the content sufficiently meets their interests and, if not, predict which page at the next level will be likely to become a more interesting one. In predicting the next-level contents, they will examine the anchor texts of various hyperlinks inside the current page. Thus, the probability, p_k, of reaching the next-level node k given the interest entropy, H_m, of entity m can be computed as follows:

$$d^*(P_m, C_k) = \begin{cases} d(P_m, C_k), & \text{if } k \in h_1, \\ \frac{H_m}{H_{max}} d(P_m, C_k), & \text{if } k \in h_2, \end{cases} \tag{6.22}$$

$$Q_j = d^*(P_m, C_j) - mean_{\forall l \in [1,h]}(d^*(P_m, C_l)), \quad j = 1, \dots, h, \tag{6.23}$$

$$U_j = \begin{cases} Q_j, & \text{if } Q_j < 0, \\ 0, & \text{if } Q_j \geq 0, \end{cases} \tag{6.24}$$

$$p_k = \frac{U_k}{\sum_{j=1}^{h} U_j}, \tag{6.25}$$

where $d^*(P_m, C_k)$ denotes the weighted distance between the preferences of entity m and the contents of node k given the entity's interest entropy H_m.

3. **Recurrent entities:** Recurrent entities are those who are familiar with the Web structure and know the whereabouts of interesting contents. They may have frequently visited such websites. Each time when they decide to forage further, they know exactly the whereabouts of the pages that closely

match their interest profiles. In this case, the probability of selecting a Web page at the next step can be defined as follows:

$$p_k = \begin{cases} 1, & \text{if } d^*(\mathbf{P}_m, \mathbf{C}_k) = min(d^*(\mathbf{P}_m, \mathbf{C}_j)), \ j = 1, \ldots, h, \\ 0, & \text{otherwise.} \end{cases} \quad (6.26)$$

6.3.3.2 Preference Updating

An entity updates its preference over time, depending on how much information on interesting topics it has found and how much it has absorbed such information. Generally speaking, the update of the preference weights in the interest profile of an entity reflects the change of its continued interest in certain topics.

When entity m reaches and finishes reading page n, it will update its interest according to the content vector of page n. The specific updating rule is defined as follows:

$$\mathbf{P}_m(\tau) = \mathbf{P}_m(\tau - 1) - \lambda \cdot \mathbf{C}_n, \quad (6.27)$$

$$pw_m^i(\tau) = 0, \ \text{for } pw_m^i(\tau) < 0, \ i = 1, \ldots, \mathbf{M}, \quad (6.28)$$

where λ denotes an absorbing factor in $[0,1]$ that implies how much information is accepted by an entity on average. $\mathbf{P}_m(\tau)$ and $\mathbf{P}_m(\tau - 1)$ denote the preference vectors of an entity after and before accessing information on page n, respectively.

6.3.3.3 Motivation and Reward Functions

As mentioned in Section 6.3.2.3, the motivational support for an entity plays an important role in information foraging. Depending on the support value, an entity will decide whether or not to forage further to the next-level Web pages. In what follows, we will describe how an information foraging entity aggregates its motivational support based on the associated motivation and reward functions.

Recall that there are three terms in Equation 6.17. The first term, $\mathbf{S}_t$, denotes the influence of initial and previously aggregated foraging support. The second term, $\Delta \mathbf{M}_t$, denotes the motivational (or patience) loss in information foraging. It changes along with the latency, i.e., the time to find information. The third term, $\Delta \mathbf{R}_t$, denotes the reward received after finding relevant information.

There are many ways to compute $\Delta \mathbf{M}_t$, which can be generally characterized as follows:

$$\Delta \mathbf{M}_t = -(\Delta \mathbf{M}_t^c + \Delta \mathbf{M}_t^v), \quad (6.29)$$

where $\Delta \mathbf{M}_t^c$ denotes the constant decrement in $\Delta \mathbf{M}_t$ at each step, and $\Delta \mathbf{M}_t^v$ denotes the variable factor that dynamically changes at each step. In our study, we propose the following models of $\Delta \mathbf{M}_t^v$:

1. As earlier studies have shown that the empirical distribution of waiting time to access Web pages follows a log-normal distribution [GVU, 2001, Helbing et al., 2000b], it is reasonable to believe that the distribution of motivational loss will be a log-normal function:

$$f_{log(\Delta M_t^v)} \sim normal(\mu_m, \sigma_m), \tag{6.30}$$

where μ_m and σ_m denote the mean and variance of the log-normal distribution of ΔM_t^v, respectively.

2. As the patience or interest of an entity in carrying on information foraging decreases as the number of required foraging steps increases, we may also adopt the following mechanism for dynamically updating the motivational loss:

$$\Delta M_t^v = \alpha_m e^{\gamma_m \cdot step}, \tag{6.31}$$

where α_m and γ_m denote the coefficient and rate of an exponential function, respectively. *step* denotes the number of pages or nodes that an entity has continuously visited.

Next, let us define the reward function in Equation 6.17. In our study, we model the reward received by an entity at each step as a function proportional to the relevant information that the entity has absorbed. In our model, since the change in the preference weights of an entity reflects the information that the entity has gained, we can write the reward function as follows:

$$\Delta R_t = \sum_{i=1}^{M}(pw_m^i(\tau - 1) - pw_m^i(\tau)). \tag{6.32}$$

Note that the reward, ΔR_t, for an entity is always greater than or equal to zero. It provides the entity with the energy to forage on the Web. On the other hand, the motivational loss, ΔM_t, of the entity is always negative, which prevents the entity to forage further. Therefore, the total support for an entity at the current step can be aggregated based on the support received at the previous steps and the changes in the above mentioned motivational loss and reward functions.

6.3.3.4   Foraging

Having defined the artificial Web space, the interest profile, and the support function of an entity, in what follows we will provide an outline of steps for simulating information foraging behavior of entities in the artificial Web space. We assume that the entities will start to forage from a homepage that contains links to other Web pages of various topics. When the support for an entity is either below a lower bound or above an upper bound, the entity will stop information foraging; otherwise, it will select and move to the next-level page. The specific steps are summarized in Algorithm 6.2.

Algorithm 6.2 The foraging algorithm.

Initialize nodes and links in artificial Web space;
Initialize information foraging entities and their interest profiles;
for each entity m **do**
    while support S $< max\_support_m$ and S $> min\_support_m$ **do**
        Find hyperlinks inside node n where entity m is visiting;
        Select, based on p_k, a hyperlink connected to a next-level page;
        Forage to the selected page;
        Update preference weights in the entity's interest profile based on Equations 6.27 and 6.28;
        Update the support function of entity m based on Equation 6.17;
    end while
    if support S $> max\_support_m$ **then**
        Entity m is satisfied with the contents and leaves the Web space;
    else
        Entity m is dissatisfied and leaves the Web space;
    end if
end for

In the next section, we will present simulated foraging results and compare them with some real-world empirical datasets for validation.

6.4. Experimentation

In this section, we will describe several experiments in which the preceding given model of information foraging entities is implemented and simulated in the artificial Web space. The objective of these experiments is to validate the entity model using some empirically obtained Web log datasets. Specifically, we want to examine whether or not the strong regularities that emerge from empirical Web log datasets can be generated in the simulations using the information foraging entities. If so, we can claim that the computational model proposed, based on the idea of information foraging, characterizes the behavior of human Web surfing that generates empirical Web regularities.

6.4.1 Experiments

In our experiments, we will apply the steps as outlined in the preceding section to initialize and control information foraging entities. As entities undertake their foraging sessions in the artificial Web space, we will record their surfing depth (steps) distribution and the rank-frequency distribution of link

Table 6.1. Parameters in Experiment 6.1.

Parameter	Value
Degree-of-coupling, r	0.7
Number of entities	5,000
Number of nodes	254
Number of topics	10
T_c	0.1
ΔM_t^c	0.2
ΔM_t^v	1st
α_u	1.5
ϕ	1.0
λ	0.6
μ_m	5.97
μ_t	1.0
σ_m	0.8
σ_p	0.25
σ_t	0.2
θ	1.0

clicks. The frequency of link clicks refers to the number of times for which a link is selected by entities. It is also called 'link-click-frequency'.

Experiment 6.1 *We initialize* 5,000 *entities foraging according to the above given motivational support and decision models for three categories of foraging entities. In this experiment, we assume that the interest profiles of the entities follow a power-law distribution and the contents of Web pages on various topics follow a normal-like distribution. Detailed experimental parameters are given in Table 6.1. We are interested in studying the distributions of entity foraging depth and link-click-frequency.*

Figures 6.1-6.4 present the statistical distributions of foraging depth and link-click-frequency as obtained in Experiment 6.1 for recurrent and rational entities, respectively.

From Figures 6.1-6.4, we can note that there do exist strong regularities in the foraging behavior of entities in the Web space. The cumulative probability distribution of entity steps in accessing pages follows a heavy-tail distribution. It is interesting to observe from Figures 6.2 and 6.4 that the distributions of link-click-frequency exhibit a power law. Similar results on the distribution of website popularity have been empirically observed and reported in [Huberman and Adamic, 1999b].

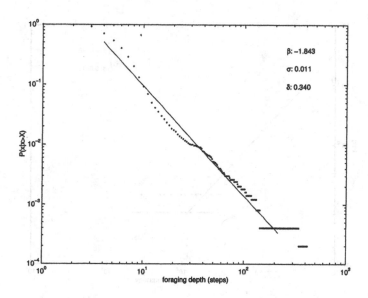

Figure 6.1. Cumulative distribution of foraging depth (steps) with recurrent entities in Experiment 6.1, where '·' corresponds to experimental data and '−' corresponds to a linear regression fitted line. The tail of the distribution follows a power-law distribution with power $\beta_c = -1.843$ and the residual of linear regression $\sigma = 0.011$. δ denotes entities' satisfaction rate (i.e., the ratio of the number of satisfied entities to the total number of entities what have surfed on the Web).

In obtaining the plots of Figures 6.1-6.4, we have applied a weighted linear regression method, in which we assign the probability at each depth or link-click-frequency with the frequency of the depth or link-click-frequency occurrence. This implies that the higher the occurrence rate of a depth or a link-click-frequency is, the higher the weight will be.

6.4.2   Validation Using Real-World Web Logs

In order to validate our model, we will use real-world Web log datasets and compare their corresponding empirical distributions with those produced by information foraging entities as mentioned above.

The first dataset is NASA Web log that recorded all HTTP requests received by the NASA Kennedy Space Center Web server in Florida from 23:59:59 August 3, 1995 to 23:59:59 August 31, 1995.[1] Before we plot the distributions, we first filter the dataset by keeping only the requests that asked for html files. This allows us to remove the noisy requests that were directly sent by

---

[1]The dataset is available from *http://ita.ee.lbl.gov/html/contrib/NASA-HTTP.html*.

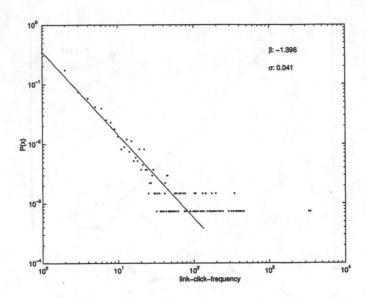

Figure 6.2. Distribution of link-click-frequency with recurrent entities in Experiment 6.1. The tail follows a power-law distribution with power $\beta_l = -1.396$, as obtained by weighted linear regression.

users, such as the requests for image files. Here, we regard a user session as a sequence of a user's continuously browsed pages on the Web, which can be derived from the filtered dataset. To obtained the user sessions, we assume that the continuous requests from the same IP correspond to the same user. We also assume that a user session ends if the idle time of a user exceeds a threshold of 30 minutes.

In the filtered NASA dataset, there are 333,471 requests in 118,252 user sessions. The average depth of surfing by users is 2.82 requests per user. In addition, there are 1,558 nodes and 20,467 links found in the dataset that were visited by users. The average links per node is around 13.

The distributions of user surfing depth and link-click-frequency for the NASA dataset are shown in Figures 6.5 and 6.6, respectively.

The second dataset is from the website of a laboratory at Georgia Institute of Technology (GIT-lab), which recorded the requests from March 26, 1997 to May 11, 1997. We preprocess the data in the same way as we did for the NASA data. As a result, we have found that there are 24,396 requests and 6,538 user sessions contained in the filtered dataset, an average of 3.73 requests per user. Also, there are 1,147 nodes and 6,984 links visited by users. The distributions of user surfing depth and link-click-frequency for the GIT-lab dataset are shown in Figures 6.7 and 6.8, respectively.

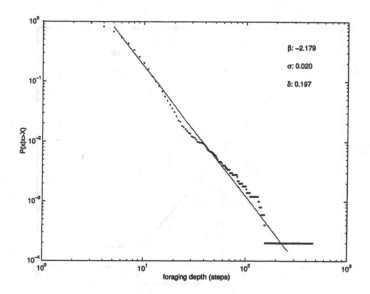

Figure 6.3. Cumulative distribution of foraging depth (steps) with rational entities in Experiment 6.1, where '·' corresponds to experimental data and '−' corresponds to a linear regression fitted line. The tail of the distribution follows a power-law distribution with power $\beta_c = -2.179$ and the regression residual $\sigma = 0.02$. δ denotes entities' satisfaction rate.

Now, let us compare the empirical distributions of Figures 6.5-6.8 with the distributions of Figures 6.1-6.4 as generated by information foraging entities in the artificial Web space. We can note that the results are quite similar, from the shapes of the distributions to the parameters of the fitted functions. The NASA dataset reveals emergent regularities closer to those produced by rational entities as in Figures 6.3 and 6.4, whereas the GIT-lab dataset presents emergent regularities closer to those produced by recurrent entities as in Figures 6.1 and 6.2. These results demonstrate that our white-box model, incorporating the behavioral characteristics of Web users with measurable factors, does exhibit the regularities as found in empirical Web log data. The foraging operations in the model correspond to the surfing operations in the real-world Web space.

In addition to the distributions of user surfing steps in accessing pages and link-click-frequency, we are also interested in the distribution of user surfing steps in accessing domains or topics − an issue of great importance that has not been studied before. We define the number of entity foraging steps in accessing domains as the number of domains that an entity has visited, and define entity satisfaction rate as the ratio of the number of satisfied entities to the total number of entities after they have completed surfing. Figures 6.9 and 6.10 present the distributions of foraging steps in accessing domains by recurrent

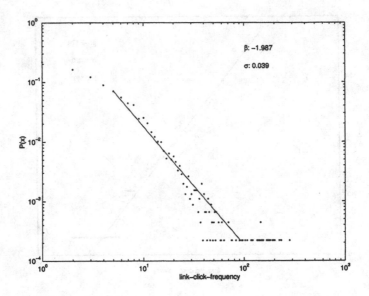

Figure 6.4. Distribution of link-click-frequency with rational entities in Experiment 6.1. The distribution follows a power-law distribution with power $\beta_l = -1.987$, as obtained by weighted linear regression.

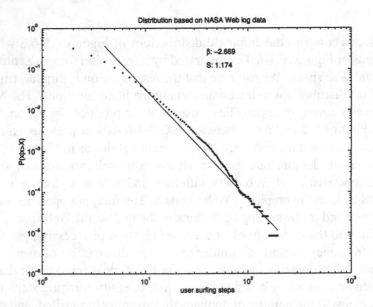

Figure 6.5. Cumulative distribution of user surfing steps as observed from the NASA Web log data. The distribution follows a heavy tail with the tail's scale of $\beta_c = -2.669$. The linear regression residual s is about 1.174.

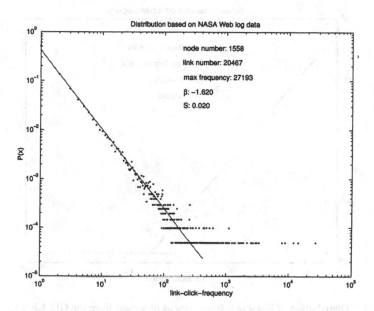

Figure 6.6. Distribution of link-click-frequency as observed from the NASA Web log data. It agrees well with a power law of power $\beta_l = -1.620$, as obtained by weighted linear regression.

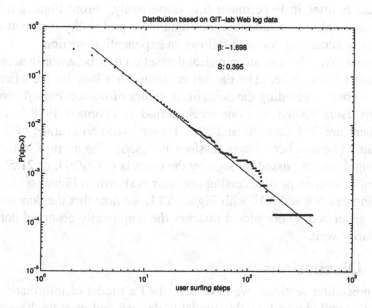

Figure 6.7. Cumulative distribution of user surfing steps as observed from the GIT-lab Web log data. The distribution exhibits a heavy tail with the tail's scale of $\beta_c = -1.698$. The linear regression residual s is about 0.395.

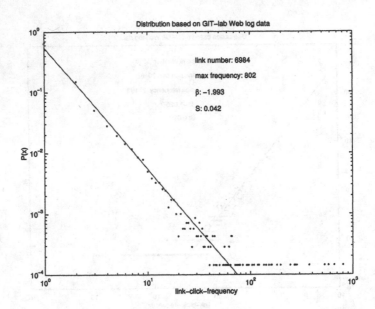

Figure 6.8. Distribution of link-click-frequency as observed from the GIT-lab Web log data. It agrees well with a power law of power $\beta_l = -1.993$, as obtained by weighted linear regression.

and rational entities in Experiment 6.1, respectively. From Figures 6.9 and 6.10, we can readily conclude that the cumulative probability distributions of entity steps in accessing domains follows an exponential function.

Let us now take a look at an empirical dataset on user behavior in accessing the domains of a website. The dataset contains Web logs for the Microsoft corporate website, recording the domains or topics of *www.microsoft.com* that anonymous users visited in a one-week period in February 1998.[2] In this dataset, there are 294 domains and 32,711 users, with an average of 3 steps per domain. The number of user sessions is 6,336. The average number of links among domains passed through by the users is 6,336/294, or 21.55. The distribution of user steps in accessing domains is shown in Figure 6.11. If we compare Figures 6.9 and 6.10 with Figure 6.11, we note that the domain-visit regularity generated by our model matches the empirically observed domain-visit regularity well.

6.5.   Discussions

In the preceding sections, we have described a model of information foraging entities and shown how this model is derived and used to characterize

---

[2]The dataset is available from *http://kdd.ics.uci.edu/databases/msweb/msweb.html*.

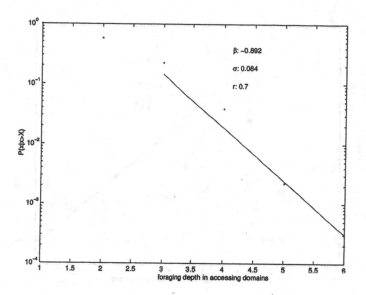

Figure 6.9. Cumulative distribution of foraging depth in accessing domains by recurrent entities in Experiment 6.1, where '·' corresponds to experimental data and '−' corresponds to a linear regression fitted line. The distribution follows an exponential function with exponent $\beta_d = -0.892$ and residual $\sigma = 0.084$.

empirically observed Web regularities. In this section, we will further investigate the inter-relationships between the emergent Web regularities as computed from our model and the characteristics of user interest profiles and content distributions.

6.5.1 Foraging Depth

One of our research objectives is to find out how the regularities in user navigation are affected by content distributions on the Web.

Experiment 6.2 *We assume that the content distribution in the Web nodes follows a power law. We are interested in examining the influence of different content distribution models on entity navigation behavior. The specific parameters for this experiment are given in Table 6.2.*

Now, let us compare the distributions of entity foraging depth in accessing Web pages as obtained from Experiments 6.1 and 6.2. Figures 6.12 and 6.13 show the foraging depth distributions of recurrent and rational entities, respectively, from Experiment 6.2. We note that the two plots in Figures 6.12 and 6.13 are almost the same as those in Figures 6.1 and 6.4, respectively. Therefore, we conclude that the regularity of entity foraging depth in accessing Web

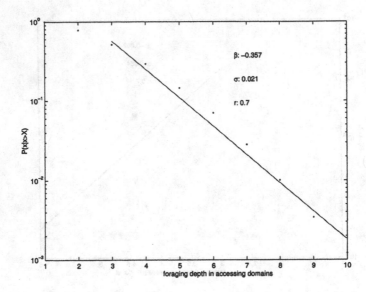

Figure 6.10. Cumulative distribution of foraging depth in accessing domains by rational entities in Experiment 6.1. The distribution follows an exponential function with a smaller exponent $\beta_d = -0.357$ and residual $\sigma = 0.021$.

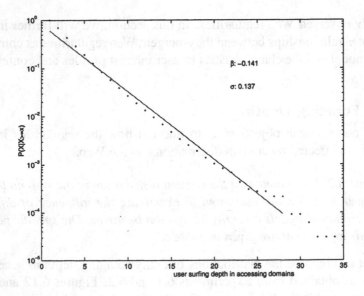

Figure 6.11. Cumulative distribution of user surfing steps in accessing domains as observed from the Microsoft Web log data. The distribution follows an exponential function with $\beta_d = -0.141$ and residual $\sigma = 0.137$.

Table 6.2.   Parameters in Experiment 6.2.

Parameter	Value
Degree-of-coupling, r	0.7
Number of entities	5,000
Number of nodes	246
Number of topics	10
T_c	0.1
ΔM_t^c	0.2
ΔM_t^v	1st
α_p	3.0
α_u	1.5
ϕ	1.0
λ	0.6
μ_m	5.97
μ_t	1.0
σ_m	0.8
σ_t	0.2
θ	1.0

pages will not be affected by the models of content distribution in the Web nodes.

Next, we will examine the effect of entity interest profiles on the Web regularities. For this purpose, we will conduct Experiment 6.3.

Experiment 6.3 *In this experiment, the interest profiles of entities are created based on a normal-distribution model. The specific parameters are given in Table 6.3. We are interested in examining the distributions of foraging depth with recurrent and rational entities, respectively.*

Figures 6.14 and 6.15 present the distributions of entity foraging depth in accessing Web pages by recurrent and rational entities, respectively, as obtained in Experiment 6.3. From Figures 6.14 and 6.15, we note that both distributions exhibit an exponential function. As the only difference between the settings of Experiments 6.1 and 6.3 is the distribution model of entity interest profiles, we suggest that the regularities of power-law distributions observed in entity foraging depth in accessing Web pages are largely resulted from the power-law distribution of entity interests in various topics.

6.5.2   Link Click Frequency

Next, let us take a look at the link-click-frequency distributions in the earlier mentioned experiments. Figures 6.16-6.19 present the distributions obtained in

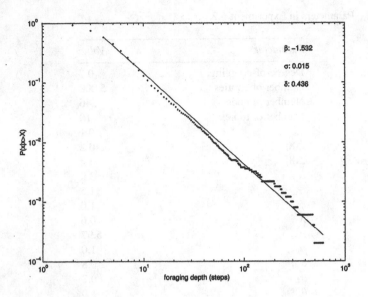

Figure 6.12. Cumulative distribution of foraging depth in accessing Web pages by recurrent entities in Experiment 6.2, where the content distribution follows a power law, different from that of Experiment 6.1. '·' corresponds to experimental data, and '−' corresponds to a linear regression fitted line. The obtained distribution follows a power law with power $\beta_c = -1.532$ and residual $\sigma = 0.015$.

Table 6.3. Parameters in Experiment 6.3.

Parameter	Value
Degree-of-coupling, r	0.7
Number of entities	5,000
Number of nodes	254
Number of topics	10
T_c	0.1
ΔM_t^c	0.2
ΔM_t^v	1st
ϕ	1.0
λ	0.6
μ_m	5.97
μ_t	1.0
σ_m	0.8
σ_p	0.25
σ_t	0.2
σ_u	0.5
θ	1.0

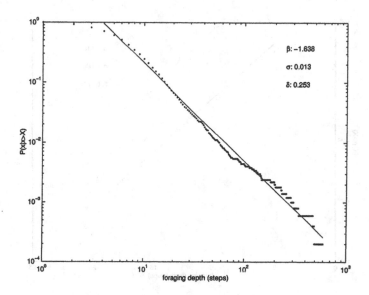

Figure 6.13. Cumulative distribution of foraging depth by rational entities in Experiment 6.2, where the content distribution follows a power law, different from that of Experiment 6.1. The distribution follows a power law with power $\beta = -1.638$ and residual $\sigma = 0.013$.

Experiments 6.2 and 6.3, respectively. As shown in the figures, the distributions of link-click-frequency remain to be a power law under the conditions of different entity interest distribution and content distribution models.

It should be pointed out that the above results can be established for recurrent and rational entities only. In the case of random entities, the regularities in link-click-frequency will disappear. Figures 6.20 and 6.21 show the plots of link-click-frequency for random entities in Experiments 6.1 and 6.2, respectively.

In fact, if we compare Figure 6.20 with Figures 6.2 and 6.4, and Figure 6.21 with Figures 6.16 and 6.17, respectively, we can observe that from random entities to recurrent entities, the power law in link-click-frequency distribution will become more and more obvious. The only distinction among the different categories of entities in our information foraging model is their ability to predict which one of the linked next-level pages may contain more interesting contents. Thus, we conclude that the power-law distribution of link-click-frequency can be affected by the content predictability of the entities.

6.5.3   Degree of Coupling

In Section 6.3.1.3, we introduced a maximum distance threshold between two linked Web pages, called degree-of-coupling, r. The larger the value of r

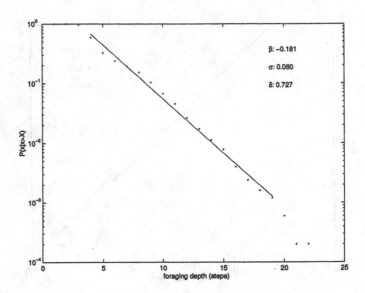

Figure 6.14. Cumulative distribution of foraging depth in accessing Web pages by recurrent entities in Experiment 6.3, where the interest profiles of entities follow a normal distribution model, different from that of Experiment 6.2. '·' corresponds to experimental data, and '—' corresponds to a linear regression fitted line. The obtained distribution follows an exponential function with exponent $\beta_c = -0.181$ and residual $\sigma = 0.08$.

is, the more links among Web pages belonging to different topics as well as the more links per each Web page. Given a certain r, the topology of the artificial Web space is determined. Entities with multiple interests will more readily forage from the contents on one topic to the contents on another topic. On the other hand, entities with a single interest will become more obsessive to select a direction from many hyperlinks on a page.

Figure 6.22 shows that the average number of links will increase as r increases. This result concerning the Web structure is commonly found on the real-world Web. The question that remains is what will be a reasonable degree-of-coupling for entities. We believe that there should be an ideal r value in our model. In order to answer this question, we will conduct Experiment 6.4 to examine the results under different r values.

Experiment 6.4 *In this experiment, we will examine the entity foraging depth and satisfaction rate with respect to r. In so doing, we will keep the rest of parameters the same as those in Experiment 6.2.*

Figures 6.23 and 6.24 show the power values in the observed power-law distributions of foraging depth and the average foraging steps, with respect to

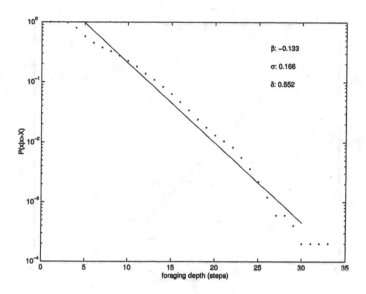

Figure 6.15. Cumulative distribution of foraging depth by rational entities in Experiment 6.3, where the interest profiles of entities follow a normal-distribution model, different from that of Experiment 6.2. The distribution follows an exponential function with exponent $\beta = -0.133$ and residual $\sigma = 0.166$.

degree-of-coupling, r, respectively. From Figure 6.23, we find that power β_c is increasing with some fluctuations. From Figure 6.24, we note that the values of average step by rational entities are higher than those of recurrent entities. The explanation for this result is that the ability to find relevant information by rational entities is weaker than that by recurrent entities, and thus rational entities must go through more pages in order to be satisfied. Consequently, their satisfaction rate will be lower than that of recurrent entities, as shown in Figure 6.25.

Website owners usually hope that visitors can stay longer or surf deeper at their websites while viewing information, and at the same time, satisfy their interests. Figure 6.26 shows the combined measure of entity foraging depth and satisfaction rate. From Figure 6.26, we observe that in order to get an optimal effect, the value of degree-of-coupling, r, should be set to $0.7 \sim 0.8$. In such a case, the average link number per node is about $11 \sim 20$, as shown in Figure 6.22.

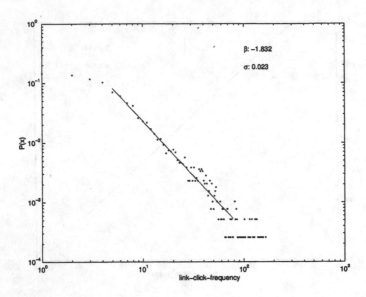

Figure 6.16. Distribution of link-click-frequency with recurrent entities in Experiment 6.2. The distribution tail is approximately a power law with power $\beta_l = -1.832$, as obtained by weighted linear regression.

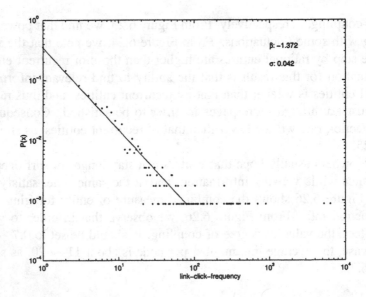

Figure 6.17. Distribution of link-click-frequency with rational entities in Experiment 6.2. The distribution is approximately a power law with power $\beta_l = -1.372$, as obtained by weighted linear regression.

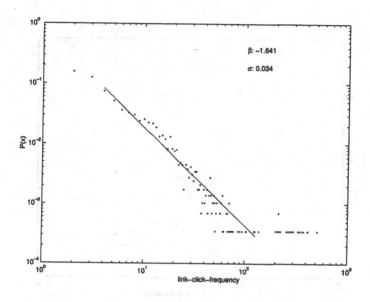

Figure 6.18. Distribution of link-click-frequency with recurrent entities in Experiment 6.3. The distribution tail is approximately a power law with power $\beta_l = -1.641$, as obtained by weighted linear regression.

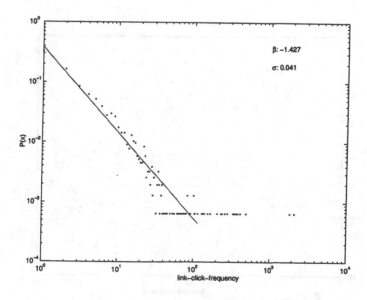

Figure 6.19. Distribution of link-click-frequency with rational entities in Experiment 6.3. The distribution is approximately a power law with power $\beta_l = -1.427$, as obtained by weighted linear regression.

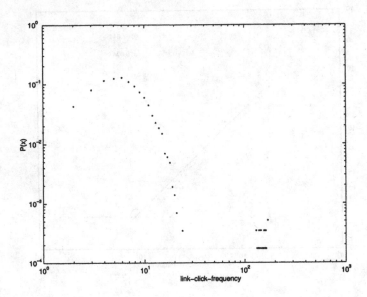

Figure 6.20.   Distribution of link-click-frequency with random entities in Experiment 6.1.

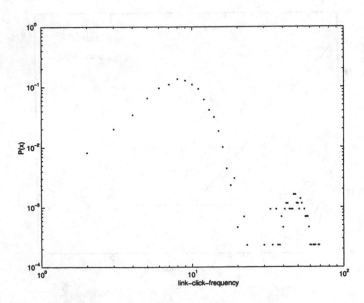

Figure 6.21.   Distribution of link-click-frequency with random entities in Experiment 6.2.

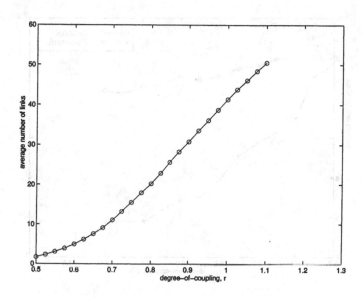

Figure 6.22. The average number of links with respect to degree-of-coupling, r, in Experiment 6.4.

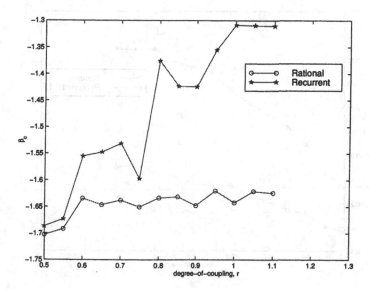

Figure 6.23. Power values, β_c, in the observed power-law distributions of foraging depth, with respect to degree-of-coupling, r, in Experiment 6.4. 'o' corresponds to rational entities and '$\star$' corresponds to recurrent entities.

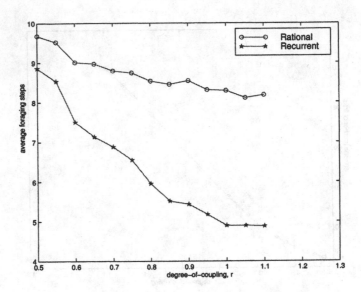

Figure 6.24. The average foraging steps, with respect to degree-of-coupling, r, in Experiment 6.4. 'o' corresponds to rational entities and '⋆' corresponds to recurrent entities.

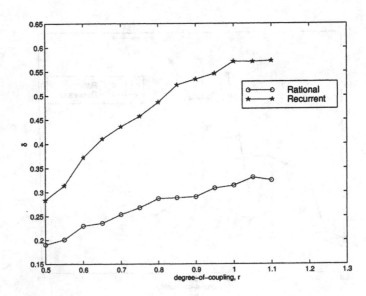

Figure 6.25. The satisfaction rate, δ, with respect to degree-of-coupling, r, in Experiment 6.4. 'o' corresponds to rational entities and '⋆' corresponds to recurrent entities.

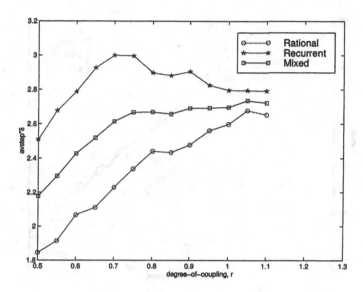

Figure 6.26. The combined measure of entity foraging depth and satisfaction rate, with respect to degree-of-coupling, r, in Experiment 6.4. 'o' corresponds to rational entities and '$\star$' corresponds to recurrent entities.

6.5.4 Mixed Entity Population

In real-world Web surfing, different users who visit a certain website can have very distinct navigation strategies. Some users may fall in the category of recurrent users, while others may be new comers. When the new comers feel that the website contains or leads to some contents of interest, they will become more likely to visit the website again. It is very important for the designer of a website to recognize from emergent Web regularities the underlying dominant navigation strategies of users.

So far, we have observed the regularities produced by three categories of information foraging entities with various interest profile distributions. It may be noted that recurrent and random entities are two extreme cases, whereas rational entities have the ability to predict the next-level contents, which is between the abilities of recurrent and random entities. The fact that all categories of users may be involved in bringing about the emergent regularities in Web surfing has led us to the following question: What will be the distributions of foraging depth and link-click-frequency if all three categories of information foraging entities are involved? In order to examine this case, we have conducted Experiment 6.5.

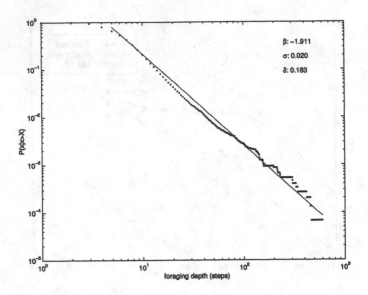

Figure 6.27. Cumulative distribution of foraging depth with a mixed population of recurrent, rational, and random entities in Experiment 6.5.

Experiment 6.5 *In this experiment, all three categories of entities, i.e., recurrent, rational, and random entities, are involved and the number of entities in each group is* 5, 000. *We are interested in examining the distributions of entity foraging depth and link-click-frequency.*

Figures 6.27-6.29 present the results of Experiment 6.5. From Figure 6.27, it can be observed that there exists a strong regularity in the distribution of foraging depth in accessing Web pages in the case of mixed entity population. The obtained result is very similar to the regularities found in empirical Web log datasets. Figure 6.28 presents the distribution of foraging depth in accessing domains, which, like the real-world statistics, follows an exponential function. Figure 6.29 shows the power-law distribution of link-click-frequency. In Figure 6.29, the occurrence point of the most probable link-click-frequency is not at 1. This is because the number of entities is too large as compared to the number of links.

To summarize, emergent regularities can readily be observed when information foraging entities make use of different navigation strategies. As far as the satisfaction rate is concerned, the mixed entity population is relatively easier to satisfy than rational entities, but more difficult than recurrent entities, as we have already shown in Figure 6.26. One way to increase the level of satisfac-

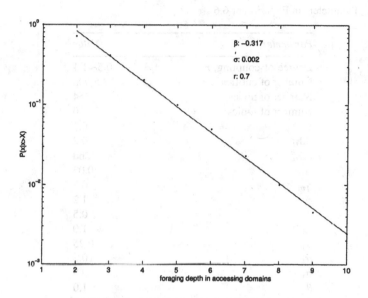

β: −0.317

σ: 0.002

r: 0.7

Figure 6.28. Cumulative distribution of foraging depth in accessing domains with a mixed population of recurrent, rational, and random entities in Experiment 6.5.

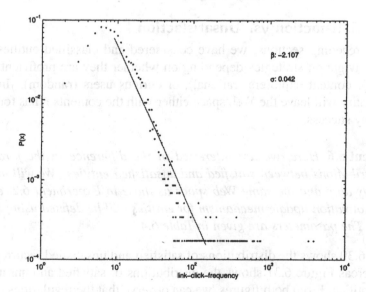

β: −2.107

σ: 0.042

Figure 6.29. Distribution of link-click-frequency with a mixed population of recurrent, rational, and random entities in Experiment 6.5.

Table 6.4.   Parameters in Experiment 6.6.

Parameter	Value
Degree-of-coupling, r	0.5~1.1
Number of entities	5,000
Number of nodes	254
Number of topics	10
T_c	0.1
ΔM_t^c	0.2
ΔM_t^v	2nd
α_m	0.03
γ_m	0.2
ϕ	1.2
λ	0.5
μ_t	1.0
σ_p	0.25
σ_t	0.2
σ_u	0.5
θ	1.0

tion rate would be to improve the descriptions of hyperlinks so that they are topic-specific and informative to foraging entities.

6.5.5   Satisfaction vs. Unsatisfaction

In the preceding sections, we have considered and classified entities with different navigation strategies depending on whether they are proficient users (recurrent), content explorers (rational), or curious users (random). In each case, an entity will leave the Web space either with the contents it has found or without any success.

Experiment 6.6 *Here, we are interested in the difference in the foraging-depth distributions between satisfied and unsatisfied entities. We will use the same entity data and the same Web space as those in Experiment 6.2, except that the motivation update mechanism for entities will be defined using Equation 6.30. The parameters are given in Table 6.4.*

Figure 6.30 shows the distributions of satisfied and unsatisfied recurrent entities, whereas Figure 6.31 shows the distributions of satisfied and unsatisfied rational entities. From both figures, we can observe that the regularities can be found in both satisfied entities and unsatisfied entities cases. This experiment also demonstrates that the regularities is not affected by the motivation update mechanism. From Figures 6.30 and 6.31, we also find that the distribution of

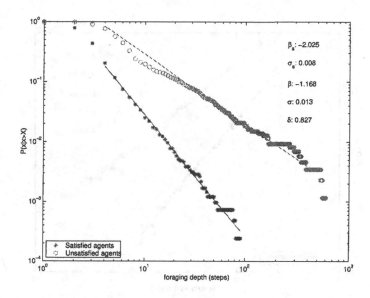

Figure 6.30. Cumulative distribution of foraging depth with recurrent entities in Experiment 6.6, with r=0.7. 'o' corresponds to unsatisfied entities and '$\star$' corresponds to satisfied entities.

unsatisfied entities has a heavier tail (i.e., higher values) than that of satisfied entities. Figures 6.32 and 6.33 present the parameter distributions in Experiment 6.6, with respect to the Web structure parameter, degree-of-coupling r.

6.5.6 Other Factors

In our study, we have conducted several other experiments to examine the possible effects on distribution regularities while changing the number of entities, the number of domains or topics, and the parameters for the distribution of entity interest profiles and for the content distribution in the Web space. The results of our experiments have consistently indicated that altering these parameters will not change the regularities of power-law or exponential distributions as mentioned above, but only alter the shape parameters for the distributions. This further indicates that the distribution regularities emerged from entity foraging behavior are stable and ubiquitous.

6.6. Summary

This chapter has presented an AOC-by-prototyping approach to characterizing Web surfing regularities. In particular, we have formulated an information foraging entity model and have validated this model against some empirical

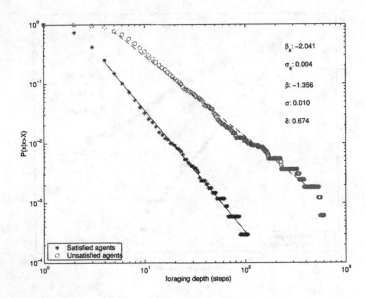

Figure 6.31. Cumulative distribution of foraging depth with rational entities in Experiment 6.6, with $r=0.7$. 'o' corresponds to unsatisfied entities and '⋆' corresponds to satisfied entities.

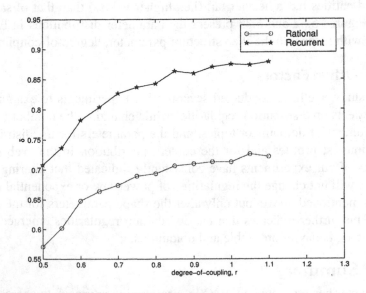

Figure 6.32. The satisfaction rate, δ, in Experiment 6.6, with r changing from 0.5 to 1.1.

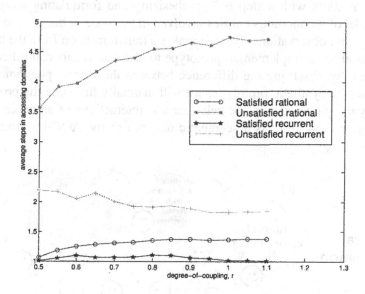

Figure 6.33. The average steps in accessing domains in Experiment 6.6, with r changing from 0.5 to 1.1.

Web log datasets. Through this example, we have shown the features of AOC-by-prototyping in complex systems modeling.

6.6.1 Remarks on Regularity Characterization

In Web-based applications, it is a common practice to record Web log data. What remains a big obstacle as well as a great challenge in Web log analysis is to characterize the underlying user behavior from the obtained data. In this chapter, we have addressed the problem of characterizing empirical Web regularities by means of a white-box, AOC model that takes into account the interest profiles, motivational support, and navigation strategies of users.

Our results have shown that based on this model, we can experimentally obtain strong regularities in Web surfing and link-click-frequency distributions. We can further examine the effects on emergent regularities after certain aspects of the Web space or the foraging behavior are changed. The given results as well as the information foraging entity based method are useful for us to understand how to develop and structure Web contents, and at the same time, how to analyze emergent user behavioral patterns.

6.6.2 Remarks on AOC by Prototyping

The AOC-by-prototyping approach is commonly applied to uncover the working mechanism behind an observed, complex phenomenon or system. In so doing, it starts with a step of hypothesizing and formulating a computational model of autonomous entities involved in the system, based on our prior knowledge and observations. Next, it makes a transformation from the hypothesized model to an implemented prototype to characterize its natural counterpart. Then, by observing the difference between the natural phenomenon or system and the synthetic prototype, we will manually fine-tune the prototype, especially the parameters in the behavior and interactions of autonomous entities. Figure 6.34 presents a schematic diagram of the AOC-by-prototyping approach.

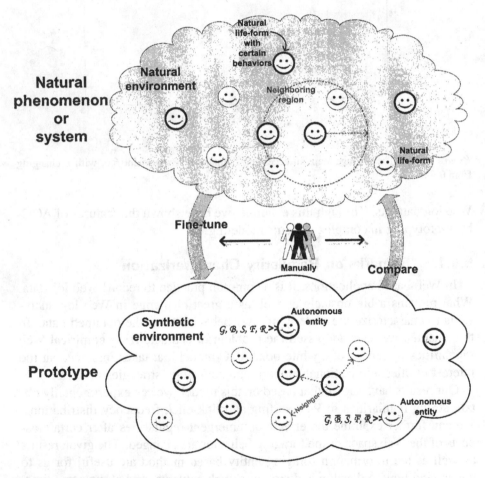

Figure 6.34. The AOC-by-prototyping approach, where the trial-and-error process, i.e., iterative fine-tune and compare steps, is manually performed (as symbolized by a human sign in the figure).

In some way, AOC-by-prototyping can be viewed as an iterative application of AOC-by-fabrication with the addition of parameter tuning at each iteration.

The difference between the desired behavior and the actual behavior of a prototype is the guideline for parameter adjustment. The process can be summarized, with reference to the summary of AOC-by-fabrication in Section 5.8.2, as follows:

1. States, evaluation functions, goals, behaviors, and behavioral rules of an entity can be changed from one prototype to the next.

2. The definition of the environment can also be changed from one version to the next. Even the model can be modified completely.

3. There is an additional step to compare the synthetic model with the natural counterpart.

4. A new prototype is built by adopting steps 1 and 2 above, and repeating the whole process.

Exercises

6.1 This chapter presents an entity-based decision model for characterizing Web surfing activities and regularities. Generalize the formulations as well as algorithms of entity decision, self-aggregated support function, and artificial Web space.

6.2 Apply the model in this chapter and in particular the support function to characterize the decision making mechanisms that lead to some self-organized regularities or phenomena as empirically observed in natural or social settings. Some examples of regularities are immune responses, traffic behavior, crowd behavior, stock markets, and computer virus attacks.

6.3 Based on the examples given in this chapter, propose and test a means of formally representing and studying the inter-relationships between emergent regularities and the characteristics of local autonomy oriented models.

6.4 Suggest and implement two strategies for tuning models as in the AOC-by-prototyping approach for a scalable application.

6.5 Perform and validate the experiments as mentioned in this chapter with empirically obtained Web client logs.

6.6 Formulate and incorporate the measurement of correlation dimension into the fine-tuning step of the AOC-by-prototyping approach.

6.7 Provide an alternative representation scheme for characterizing generic Web contents (e.g., Web services).

6.8 The results obtained from an AOC-by-prototyping study can offer further insights into new models and formulations for AOC-by-fabrication solutions to certain computational problems. Explain how these results (e.g., immune response characterization and information foraging modeling) may lead to more powerful adaptive computation and distributed optimal search techniques, respectively.

Chapter 7

AOC in Optimization

7.1. Introduction

Based on the AOC-by-fabrication and AOC-by-prototyping techniques described in the previous chapters, we can start finding a complex system in nature after which we can model and build an AOC-based problem solver. This process is usually an iterative one requiring a certain amount of trial-and-error effort. This is certainly not an easy task, especially for optimization problems with many factors (or dimensions) to consider.

AOC-by-self-discovery emphasizes the ability of AOC to find its own way to achieve what AOC-by-prototyping can do. The ultimate goal is to have a fully automated algorithm that can adjust its own parameters for different application domains. In other words, the AOC itself becomes autonomous.

To illustrate the AOC-by-self-discovery approach, in this chapter we will describe a population-based stochastic search algorithm, called evolutionary diffusion optimization (EDO), as inspired by diffusion in nature. We will show in detail the process of developing EDO. Below is some background information to help readers understand optimization, some common algorithms for solving optimization problems, and the diffusion model that inspired EDO. A full description of EDO with an emphasis on the self-adaptive behavior will be given, together with a discussion on the behavior and problem solving power of EDO.

7.1.1 Optimization Tasks

Real-world applications of optimization can be found in the fields of manufacturing, logistics, finance, bioinformatics, transportation system control, spacecraft trajectory optimization, VLSI layout design, etc. To illustrate their characteristic features, two examples in financial engineering and telecommunication network design are described below.

7.1.1.1   Investment Portfolio Optimization

The primary objective of investors is to maximize their returns on investment. Each stock has its own risk and returns. To obtain a bigger return, an investor should, therefore, be willing to bear a higher risk. However, the willingness of an investor to take risks decreases as the earning increases. Even if it is irrespective of the risk tolerance of an investor, a good feeling on the market trend is required before any investment can yield a positive return.

Common sense tells us "never put all the eggs in one basket." An investor usually diversifies its investment in several stocks. In fact, this is a portfolio selection problem that can be formulated into an optimization problem as follows [Perold, 1984]:

Let w_i be the weight of stock i that has an expected return r_i. Then, the expected return of a portfolio, z, of n stocks is

$$E(z) = \sum_{i=1}^{n} w_i r_i. \tag{7.1}$$

It can be written as

$$E(z) = r^T w, \tag{7.2}$$

where $r = \langle r_1, r_2, \cdots, r_i, \cdots, r_n \rangle$ is the expected return vector and $w = \langle w_1, w_2, \cdots, w_i, \cdots, w_n \rangle$ is the weight vector.

Correspondingly, the covariance matrix, Q, of the expected return can be written as follows:

$$Q(r) = \sum_{i=1}^{n} \sum_{j=1}^{n} (r_i - E(r_i))(r_j - E(r_j)). \tag{7.3}$$

Usually, an investor's aversion to risk is represented by a utility function, $f(x) = 1 - e^{-kx}$, where $k > 0$ is the risk aversion parameter.

Given a covariance matrix, Q, a return vector of expected returns, r, and a risk aversion parameter, k, then the above portfolio selection problem can be formulated as the following maximization problem by considering the expected return and its corresponding risk [Perold, 1984]:

$$
\begin{aligned}
\max \quad & r^T w - \tfrac{k}{2} w^T Q w, \\
\text{s.t.,} \quad & \sum_{i=1}^{n} w_i = 1, \\
& w_i \leq 0.
\end{aligned}
\tag{7.4}
$$

In essence, the optimal portfolio is determined by the weighting parameter, w. Interested readers are referred to [Perold, 1984] for more details on large-scale portfolio optimization.

7.1.1.2 Telecommunication Network Design

A telecommunication network connects base stations on different positions together where the communication equipment, such as host computers, concentrators, routers, and terminals, is located. Designing a telecommunication network is to find the most efficient way to connect the base stations. This problem is sometimes called a minimum spanning tree problem. In the following, we will formulate it into an optimization problem.

Given a group of base stations, $V = \{1, 2, \cdots, n\}$, and the cost, c_{ij}, for connecting two stations i and j, the problem is actually to determine a graph $G = \langle V, E \rangle$, where $E = \{\langle i, j \rangle \mid i, j \in V\}$, as follows:

$$\begin{aligned} \min \quad & \textstyle\sum_{i=1}^{n-1} \sum_{j=2}^{n} c_{ij} x_{ij}, \\ \text{s.t.,} \quad & \textstyle\sum_{i=1}^{n-1} \sum_{j=2}^{n} x_{ij} = n - 1, \\ & \textstyle\sum_{j=1}^{n} x_{ij} < d_i, i \in V, \\ & x_{ij} = 0 \text{ or } 1, i, j \in V, \end{aligned} \tag{7.5}$$

where d_i is the upper bound of the number of links connected to base station i, and

$$x_{ij} = \begin{cases} 1, & \text{if link } \langle i, j \rangle \text{ is selected in a spanning tree,} \\ 0, & \text{otherwise.} \end{cases} \tag{7.6}$$

More details on this and similar problems can be found in [Gen et al., 2000]. To solve the minimum spanning tree problem by a population-based algorithm, it is appropriate to encode the tree structure within an autonomous entity. Possible autonomous behaviors include pruning one's own tree or exchanging a partial tree with another autonomous entity.

7.1.2 Objectives

As mentioned previously, the goal of an optimization task can be represented mathematically. Consider the goal being written as a function $F(x)$ where $x = \{x_1, x_2, \cdots, x_n\}^T$ is an n-dimensional vector representing the parameters of function F. The optimal solution is represented by $F(x^*)$ such that

$$\forall\, x, \; F(x^*) \leq F(x). \tag{7.7}$$

The search for x^* can be viewed as the minimization of function F. Turning the sign in Equation 7.7 around makes the search for x^* a maximization task. They can collectively be called global optimization tasks [Torn and Zilinskas, 1989].

There are several challenges any search algorithm must face. First, the landscape of the function to be optimized is unknown. Unimodal functions can be monotonic in nature and the search is easy once the downhill direction is found. However, finding the direction of the search landscape is not a simple task. Multimodal functions, on the other hand, have many suboptimal solutions where a search algorithm is likely to be trapped.

Secondly, there is usually no linear relationship between changes made to the function variables and the corresponding change in the function value. This credit assignment problem confuses, if not misleads, the search algorithm.

Thirdly, search algorithms do not normally jump directly to the optimal solution, but make incremental changes in small steps instead. Making big steps is not always a better strategy, especially when the optimal solution is close by. In contrast, infinitesimal changes are detrimental to the effort to escape from a local optimum. Therefore, it is crucial to choose a step size appropriate to the situation prevalent during the search.

Finally, a population-based search algorithm needs to maintain a sufficient diversity during the whole course of the search so that the search space is adequately sampled.

Many algorithms have been developed over the years to tackle the challenging task of global optimization [Horst and Pardalos, 1995, Horst and Tuy, 1990, Mockus, 1989]. In the absence of prior knowledge about the search landscape, stochastic methods, such as simulated annealing [Kirkpatrick et al., 1983] and population-based incremental learning [Baluja, 1994, Baluja and Caruana, 1995], have been proven to be effective. They attempt to locate the optimal solution by generating sampling points probabilistically. Methods inspired by nature that are equally successful include evolutionary algorithms [Bäck et al., 1997], bacterial chemotaxis [Müller et al., 2002], and differential evolution [Storn and Price, 1997].

In order to demonstrate the AOC-by-self-discovery approach in optimization problem solving, a new population-based stochastic search algorithm, called evolutionary diffusion optimization (EDO), will be presented in this chapter. Unlike the population-based algorithms mentioned above, the population size in EDO is not fixed because it is very lenient towards poor performers in the population. This is a major departure from the normal practice of many nature inspired algorithms, such as evolutionary algorithms, particle swarm optimization [Kennedy, 1997], cultural algorithm [Reynolds, 1994], and ant colony optimization [Dorigo et al., 1996]. Moreover, candidate solutions are changed based on landscape information gathered by entities in the same family. The right to reproduce is also granted based on local competition. We will give a more detailed discussion on the commonalities and differences between EDO and the above nature inspired algorithms later in Section 7.7.2.

The AOC-based method, EDO, is specifically designed with the following objectives:

1. To learn the search landscape by group efforts and information sharing;

2. To maintain a high population diversity;

3. To automatically adjust search step sizes.

7.2.   Background

Before we go into details on the proposed AOC-by-self-discovery approach to global optimization, we will describe below two well-known stochastic search algorithms as well as diffusion in natural systems, which are the inspiration sources of our EDO method.

7.2.1   Common Stochastic Search Algorithms

To find the optimal solution, different methods have been proposed. There are methods that look for exact solutions to a given optimization problem by performing some procedures iteratively with a set of deterministic behaviors. There are also methods that behave in a non-deterministic or stochastic way. In the context of AOC algorithms, the latter is more relevant to the discussion and useful for benchmarking purposes.

7.2.1.1   Simulated Annealing

Simulated annealing is a stochastic search algorithm that mimics the annealing procedure in physics. When applied to optimization problems, the starting point is randomly generated. Subsequent sampling points are chosen by making random changes to a current point. A new point is accepted if its function value is not greater than the point at hand. Otherwise, it will accepted probabilistically. The difference between the function values of the current and the new point, together with a gradually decreasing temperature parameter, determines the chance of a worse point being accepted. The temperature parameter also affects the amount of changes at each iteration. Hence, changes at the beginning of the search are relatively bigger than those at latter stages. The diversity in the samples generated is maintained by the point generation function. The crucial factor to success is the rate at which the temperature parameter is decreased, which is problem dependent. Adaptive simulated annealing [Ingber, 1996] tackles this problem by periodically adjusting the bounds of the temperature based on the system performance.

7.2.1.2   Evolutionary Algorithms

Evolutionary algorithms (EA) exist in many different forms, such as genetic algorithms (GA) [Holland, 1992], evolutionary programming (EP) [Fogel et al., 1966], and evolution strategies (ES) [Schwefel, 1995]. Recently, fast evolutionary programming (FEP) [Yao et al., 1999] has been proposed as an improvement over the canonical evolutionary programming and has been shown to be superior in optimizing a large number of functions. The mutation operator makes changes to the function variable based on a randomly chosen value from a Cauchy function, which is similar to a Gaussian function with zero mean but with a wider spread. Therefore, there is a high probability of making zero and small changes with occasionally large changes. The adoption of the Cauchy function has taken care of the step size issue effectively. In addition, FEP adapts the spread of mutations over time. Hence, adding further adaptation to the search step size. Mutation also helps maintain the population diversity. The selection process in FEP tackles the credit assignment problem by admitting better performers to the next generation. The same Cauchy function has also been used in the fast evolution strategies (FES) algorithm to improve the performance of its canonical counterpart [Yao and Liu, 1997].

7.2.1.3   Adaptive Evolutionary Algorithms

Adaptive evolutionary algorithms [Angeline, 1995, Bäck, 1997, Grefenstette, 1986, Hinterding et al., 1997] have been proposed that automatically tune some of the parameters related to the algorithms, such as mutation rate [Bäck, 1992, Bäck and Schutz, 1996, Gzickman and Sycara, 1996, Schwefel, 1981, Sebag and Schoenauer, 1996, Williams and Crossley, 1997], mutation operator [Liang et al., 1998, Michalewicz, 1994, Swain and Morris, 2000, Yao et al., 1999], crossover rate [Ko and Garcia, 1995, Smith and Fogarty, 1996], crossover operator [Angeline, 1996, Ko et al., 1996], and population size [Williams and Crossley, 1997]. They usually track changes in progress measures, such as online and offline performance [Grefenstette, 1986], the ratio of average fitness to the best fitness, and the ratio of the worst fitness to average fitness, among others.

Meta-EA is another group of self-improving EA that does not rely on the specific instruction of the designer [Freisleben, 1997]. An EA [Bäck, 1994, DeJong, 1975, Freisleben and Härtfelder, 1993a, Freisleben and Härtfelder, 1993b] or another intelligent system [Chung and Reynolds, 2000, Herrera and Lozano, 1998, Lee and Takagi, 1993, Tettamanzi, 1995] is used to control a population of EA in the way similar to an EA optimizing the parameters of the problem at hand.

Other examples of adaptive algorithms include the evolution of emergent computational behavior by employing a GA to evolve the rules of a cellular automaton for a synchronization task [Crutchfield and Mitchell, 1995, Das et al., 1995, Hordijk et al., 1998] and for generating test patterns for hardware circuits [Chiusano et al., 1997]. A variant of the latter example employs a selfish gene algorithm [Corno et al., 2000].

7.2.2   Diffusion in Natural Systems

Diffusion in nature and the successful application of the diffusion models to image segmentation (see Section 2.3.2) have inspired the evolutionary diffusion optimization (EDO) algorithm, which attempts to tackle the task of optimizing multi-dimensional functions. Specifically, building on these models, EDO is designed to include strategies to learn the trend of the search space, while forming chains of information flow to facilitate the small step exploration of the search space. This section describes in more detail the diffusion models observable from natural systems.

The study of human geography categorized reasons for human migration into push factors and pull factors [Bednarz, 2000, Clark, 1986]. Push factors relate to undesirable conditions, such as poor living conditions, lack of jobs and overcrowding. Pull factors are those positive factors that attract people to relocate, such as jobs and better living conditions. Two forms of migration can also be identified. Step migration refers to a series of local movements, such as moving from village to town, then to city. Chain migration refers to a more drastic change beyond the local region and is usually assisted by people who have already emigrated. The availability of information seems to be an important factor that helps people decide when and where to migrate.

The important lesson to learn for function optimization is that once the landscape of the function to be optimized is known, finding the optimal solution becomes trivial. The question is how to capture the trend. A search algorithm will need past experiences to inform it of the possible unsuccessful moves (push factor) and successful moves (pull factor).

In addition, knowledgeable people can help others decide whether to move or not. In essence, whoever has captured the trend of the search space can help others make better moves by sharing their knowledge. On the other hand, as the human migration model shows, the chance of finding a solution can possibly be enhanced by making small step migrations.

7.3. EDO Model

The EDO model maintains a population of autonomous entities that will decide for themselves what to do regarding the search. A set of primitive behaviors have been defined in the evolutionary model. Algorithm 7.1 shows the major steps in EDO.

Entities in EDO are divided into two categories: active and inactive. Active entities are those that perform the search task for the optimal solution. They are engaged in diffusion behavior and negative feedback. In contrast, inactive entities are those that have found positions better than those of their parents, which may not necessarily correspond to suboptimal or global optimal solutions. These entities will not diffuse any more, but perform reproduction behavior and give positive feedback again to their parents. EDO will control the population size via the aging behavior of entities. It will also adapt some global parameters. The following sections detail the primitive behaviors of entities and several global parameters in an EDO system.

Algorithm 7.1 The EDO algorithm.

---

while (the number of entities > 0) and (the number of iterations < limit) **do**
    Evaluate an entity;
    if the entity performs better than its parent **then**
        Get positive feedback;
        Reproduce;
        Its parent becomes inactive;
    else
        Get negative feedback;
        Diffuse;
        Age;
    end if
end while

Procedure *Diffuse*
if rand() > P_{rand_move} **then**
    Random-move;
else
    for each variable **do**
        Select its step direction and size;
    end for
end if

Procedure *Reproduce*
quota ← $f(fitness)$;
Create a new probability matrix;
for each offspring **do**
    Copy variables and point to the new probability matrix;
    Diffuse;
end for

Procedure *Age*
age ← age +1;
if (its fitness < its parent's fitness × threshold) or
((age > lifespan) and (its fitness < the average value)) **then**
    Remove;
end if

7.3.1   Diffusion

Entities in EDO explore uncharted positions in the solution space by performing their diffusion behaviors. A diffusion behavior is an operation through which an entity modifies its object vector, which is a set of values corresponding to the variables of the function to be optimized. The object vector **V** can be represented as follows:

$$V = \{v_1, v_2, \ldots, v_n\}, \ \forall \, i, \ v_i = [LB, UB], \tag{7.8}$$

where LB and UB are the lower and upper bounds, respectively. All function variables can take values within these bounds.

Entities in EDO have two types of diffusion behavior: rational-move and random-move. In both cases, the updated object vector becomes the new position of the entity in the solution space. However, if an entity remains stationary after diffusion (i.e., it fails to diffuse and stays at its current position), the process will be repeated.

- **Rational-move:** In the majority of time, an entity performs rational-move behaviors. In so doing, it modifies its object variables by randomly choosing the number of steps to take, according to a probability matrix (see Section 7.3.4). The actual amount of change is the product of the current step size and the number of steps chosen:

$$\forall \, i, \ v_i = v_i + \delta v_i \cdot \Delta, \tag{7.9}$$

$$\delta v_i = \min\{k \mid \text{rand}() < \sum_{j=-y}^{k} p_{i,j}, \ k \le y\}, \tag{7.10}$$

where v_i is the ith function variable, δv_i is the number of steps to be taken, Δ is the step size, y is the maximum number of allowable steps towards either end of the bounds in the search space, and $p_{i,j}$ is the probability of v_i making j step(s).

- **Random-move:** As an entity becomes older and has still not located a position better than that of its parent, it will decide to perform a random-move behavior with an increasing probability. The probability, P_{rm}, to perform a random-move is given by:

$$P_{rm} = exp\left[-\frac{\Theta - a}{\alpha}\right], \tag{7.11}$$

where α is a scaling factor that decides the degree to which the random-move is to be exercised, Θ is the maximum lifespan of an entity, and a is the age of an entity.

The direction of movement is first chosen uniformly between towards the upper bound, towards the lower bound, and no move. In the case that a move is to be made, a new value between the chosen bound and the current value is then randomly selected. Specifically,

$$\forall\, i,\ v_i = v_i + \text{rand}() \cdot (r_{<l>} - v_i), \qquad (7.12)$$

$$l = \min\{k \mid \text{rand}() < \sum_{j=1}^{k} b_j,\ k \leq 3\}, \qquad (7.13)$$

$$r = \{LB, v_i, UB\},\ \forall\, j,\ s_j = 1/3, \qquad (7.14)$$

where r is the set of boundaries between which a new value will be chosen for object variable v_i, and s is the set of probabilities for choosing the entries in r.

7.3.2   Reproduction

At the end of an iteration in EDO, the fitness values of all active entities are compared with those of their parents, which have temporarily become stationary. All entities with higher fitness values will perform a primitive behavior, called reproduction. A reproducing entity will replicate itself a number of times as decided by a quota system. The offspring entities are then sent off to new positions by rational-moves.

Fitness, f, in EDO measures an entity's degree of success in the course of the search for the optimal solution. It is also used as a basis to determine various primitive behaviors. For simplicity, the objective function value is used as fitness in EDO if the task is minimization. The reciprocal of the function value can be used as fitness if EDO is used in a maximization task.

7.3.2.1   Reproduction Quota

The number of offspring entities to be reproduced, i.e., quota q, is governed by the entity's fitness and two system-wide parameters: maximum offspring, Ω, and maximum population size, Π. The following two rules are applied in succession:

Differentiation Rule.   An entity is given the full quota to reproduce only if its fitness is significantly above the population average and will gradually decrease as the fitness decreases. Therefore, the quota for an entity e having fitness f is:

$$q_x = \begin{cases} \Omega, & \text{if } \ \frac{f_e}{\overline{f}} \leq \omega_1, \\ \Omega - 1, & \text{if } \ \omega_1 < \frac{f_e}{\overline{f}} \leq \omega_2, \\ \Omega - 2, & \text{otherwise,} \end{cases} \qquad (7.15)$$

where $\overline{f}$ is the population average fitness, and ω_1 and ω_2 are the intervals in the step function.

Population Size Rule. Further to the above differentiation scheme, the reproduction quota is subjected to a further restriction to avoid overcrowding:

$$q_x = \lfloor \frac{q_x * (\Pi - \Sigma)}{\Pi} \rfloor, \tag{7.16}$$

where Σ is the size of the entity population, and Π is the maximum population size.

7.3.2.2 Rejuvenation

An inactive entity will be allowed to perform a primitive behavior, called rejuvenation, in order to spawn new offspring if the following two conditions are satisfied:

- All its offspring entities are dead.

- Its fitness is better than the population average.

The main idea behind the rejuvenation behavior is that an inactive entity has been receiving positive or negative reinforcement signals from its offspring (described below), its probability matrix contains the latest information regarding the neighborhood landscape. It would be a waste if this potentially useful information is discarded. A rejuvenated parent will be given the full quota, Ω, and then subjected to the population size rule (Equation 7.16) to reproduce its offspring.

7.3.3 Aging

Through the aging behavior of entities, EDO keeps track of the unproductive moves throughout the search. By limiting the number of unsuccessful moves, EDO can properly channel resources to explore the search space. However, sufficient time should be put aside to allow each entity to survey its neighborhood.

The age, a, of an entity in EDO denotes the number of iterations for which this entity has survived since its birth. Once an entity becomes a parent, the age does not need to be updated any more. All entities in EDO will only be allowed to perform search for a certain number of iterations, because we do not want to have too many non-contributing entities in the system. The global lifespan information, Θ, is the maximum number of allowable iterations for which any entity can survive.

At the end of each iteration, all entities perform an aging behavior to increase their ages by one. Entities whose age is greater than the lifespan limit are eliminated from the system. However, it is necessary to provide exceptions to this rule to either retain the exceptionally good performer, or prematurely eliminate the exceptionally poor ones.

- **Extended life:** The lifespan limit of an entity is extended by one iteration when it expires, if its fitness is higher than the population average.

- **Sudden death:** An unsuccessful entity will be eliminated, if its fitness is less than a certain percentage of the fitness of its parent. The threshold is set at 80% in the experiments reported later.

7.3.4 Feedback

Each entity in EDO performs an information-passing behavior to pass information back to its parent, if it has moved to a better or worse position. This information allows the parent to update its probability matrix, which is shared among its offspring.

A probability matrix, p, contains the likelihood estimate of success with respect to the direction of a move. Specifically, the probability matrix is an $n \times m$ matrix representing n function variables to be optimized and m possible steps (including $y = (m - 1)/2$ steps towards the upper bound and the lower bound, respectively, and the current position). A global step size parameter, Δ, governs the unit of change in all function variables. The product of Δ and the number of steps becomes the final modification to affect on V. Formally,

$$p = \{p_1, p_2, \ldots, p_n\}, \tag{7.17}$$

$$p_i = \{p_{i,-y}, \ldots, p_{i,0}, \ldots, p_{i,y}\}, 0 \le p_{i,j} \le 1, \tag{7.18}$$

$$\forall i, \ \sum_{j=-y}^{y} p_{i,j} = 1. \tag{7.19}$$

At the beginning of the search, all entries are initialized to $1/m$, which means they are equally likely to make any of the possible moves. The probability matrix will be continuously updated as a result of the local feedback behavior of an entity. As the probabilities for various steps are updated, EDO begins to differentiate between the good and bad directions of moves with respect to the initial position of the ancestor. The probability matrix, therefore, facilitates information sharing between siblings.

Sharing this information is made possible by having all offspring entities of the same parent use the same probability matrix. The motivation behind this is that a trend is a kind of local information, it will become increasingly irrelevant to places further and further away from that of a parent entity. Hence, it is necessary to update the probability matrix using new local information.

- **Positive feedback:** A successful move, which may happen after taking many moves, is defined as a gain in fitness. In order to bias the future moves of an entity's siblings to its own successful move, we update the probabilities in the parent's probability matrix, which correspond to the changes made in the successful entity's object vector. The updating rule is as follows:

$$p_{i,j} = \frac{p_{i,j} + \beta}{1 + \beta},$$
(7.20)

where $p_{i,j}$ is the probability that relates to the ith function variable and jth step size, and β is the learning rate.

- **Negative feedback:** In order to steer the siblings of an entity away from a non-optimal area in the search space, an entity will update the probability matrix of its parent after each unsuccessful move, using the following rule:

$$p_{i,j} = p_{i,j} \cdot (1 - \beta),$$
(7.21)

where β is the same learning rate as used in positive feedback.

As compared to positive feedback, negative feedback is a finer grain update as it happens after each step. Moreover, the use of a multiplicative scaling factor ensures that the probability remains greater than zero at all time. The whole probability matrix is normalized after updating within the set of probabilities for each dimension.

In summary, an entity, e, in the population, P, maintained by EDO is a tuple (V, p, a, f), where V is the object vector, p is the probability matrix, a and f are scalars representing the age and fitness of entity e, respectively. While V contains the values of the potential solution, p, a, and f are crucial to the search process of EDO.

7.3.5   Global Information

Various system-wide parameters have been mentioned in the previous sections. They either control the process of the search or act as parameters to some of the features in EDO. Below is a summary of the system-wide parameters:

1. Step size (Δ): It is the size of a step to be taken during a rational-move. It is used in combination with the probability matrix to decide what the actual change should be. For example, if two steps towards the upper limit were chosen based on the probability distribution in the probability matrix, a value equal to two times of Δ will be added to the function variable in question.

2. Lifespan (Θ): This is the duration, in iterations, given to an entity to perform a search. It aims at limiting the amount of unsuccessful exploration that any entity can perform before it is eliminated.

3. Maximum offspring (Ω): This is the maximum number of offspring any reproducing entity is allowed to spawn at a time. It represents the amount

of local exploration an entity is allowed to perform in the neighborhood of a good solution.

4. Maximum population size (Π): This is the upper limit of the number of entities, both active and inactive, an EDO system can keep at any time.

EDO has some adaptive behaviors to assist the optimization process. First, the step size parameter, Δ, is reduced according to the golden ratio[1] if the current best solution has not been renewed for half of lifespan, Θ. The rationale behind this reduction is that the entities may be in the neighborhood of a minimum. Therefore, finer steps are required for careful exploitation. Conversely, if the population has been improving continuously for some time (in number of iterations, again equal to half of lifespan), the step size is increased according to the golden ratio (division):

$$\Delta = \begin{cases} \Delta * \phi, & \text{if } u < \Theta/2, \phi = \frac{1+\sqrt{5}}{2}, \\ \Delta/\phi, & \text{otherwise,} \end{cases} \qquad (7.22)$$

where u is the number of times the current best solution has been renewed since the step size parameter was last updated, and ϕ is the golden ratio.

EDO also reduces the maximum population size by a small amount at each iteration to increase the selection pressure.

7.4.   Benchmark Optimization Problems

Four benchmark functions are chosen to test the EDO algorithm (see references within [Yao and Liu, 1997, Yao et al., 1999]). f_1 and f_2 are unimodal functions, while f_3 and f_4 are multimodal functions:

$$f_1(x) = \sum_{i=1}^{n} x_i^2, \qquad (7.23)$$

$$f_2(x) = \sum_{i=1}^{n} \left(\sum_{j=1}^{i} x_j \right)^2, \qquad (7.24)$$

$$f_3(x) = \sum_{i=1}^{n} [x_i^2 - 10\cos(2\pi x_i) + 10)], \qquad (7.25)$$

$$f_4(x) = \left[\frac{1}{500} + \sum_{j=1}^{25} \frac{1}{j + \sum_{i=1}^{2}(x_i - a_{ij})^6} \right]^{-1}, \qquad (7.26)$$

---

[1]The golden ratio is one of the roots to the equation $x^2 - x - 1 = 0$, and many visually appealing geometrical shapes in nature have a golden ratio dimension [Dunlap, 1997].

where

$$a_{ij} = \begin{pmatrix} -32 & -16 & 0 & 16 & 32 & -32 & \cdots & 0 & 16 & 32 \\ -32 & -32 & -32 & -32 & -32 & -16 & \cdots & 32 & 32 & 32 \end{pmatrix} \qquad (7.27)$$

The unimodal functions are used to demonstrate the hill climbing capability of EDO and the multimodal functions are used to test EDO's ability to escape from local minima.

7.5. Performance of EDO

Using the most basic (two-dimensional) form of f_1 (Equation 7.23) and f_4 (Equation 7.26) as test cases, this section will present the performance of EDO as outlined above.

The basic form of EDO consists of primitive behaviors, such as diffusion, reproduction, aging, and feedback. In order to show the effects of some new features, namely random-move during diffusion (Equation 7.12), step size adaptation (Equation 7.22), and solution rejuvenation, various combinations of the basic EDO with these features are used to optimize the two-dimensional version of f_1 (Equation 7.23) for 200 iterations. Figure 7.1 shows the best and average function values with the basic EDO together with one of the three new features. Figure 7.2 shows the plots of the same information for the combination of the basic EDO with more than one of the features.

Basic EDO for a Good Solution. The basic EDO is able to find a good enough solution, which has a value less than 1, in just 10 iterations (see Figure 7.1(a)). This is a period when the diversity in terms of the variety in the parameter values is at the highest. The population converges gradually during this period. As the diversity decreases, EDO tends to try out small variations to the candidate solutions in the population, and attempts to update the probability matrix. As the population moves together closer to the optimal solution, a fixed step size will gradually move the candidate solutions across the optimal point and over to the other side of the valley in the function value landscape. In combination with the fact that EDO accepts all new points into the solution pool, the population average will diverge at a certain point. In addition, whatever direction bias information learned in the probability matrix will suddenly become irrelevant. A new learning phase then commences (between iterations 40-50). This cycle of learning and re-learning can also be observed at the later stages of the search (between iterations 100 and 200).

Another point worth noting is that the new best solution is not discovered at every iteration. One may expect that once a trend has been learned and encoded into the probability matrix, any new entity having the latest probability matrix will simply slide downward towards the optimal solution. This is not found in any of the plots in Figure 7.1, because new candidate solutions are generated using the probability matrix and the step size. However, this fixed size is be-

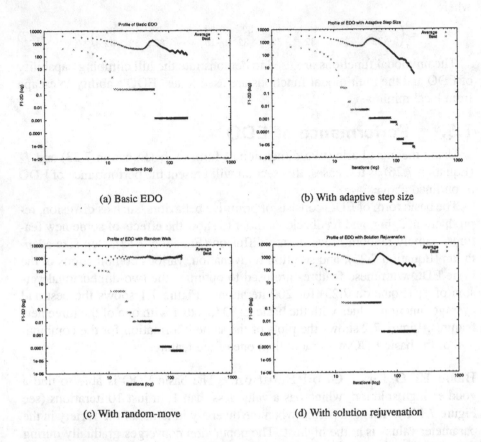

(a) Basic EDO   (b) With adaptive step size

(c) With random-move   (d) With solution rejuvenation

Figure 7.1.   The changes in the best and average values of the population when EDO searches for a solution to the two-dimensional unimodal function f_1. (a) shows the result for the basic EDO, which has the basic diffusion, reproduction, aging, and feedback mechanisms; (b) the adaptive step size strategy modifies how far one step is in EDO, and gives the best result with a converging population; (c) random-move, which allows the entities to make big jumps instead of relying on the probability matrix for direction selection, gives better results than the basic EDO, but the population diverges; (d) solution rejuvenation, which allows the above average candidate solution to reproduce when all its offspring entities are dead, does not show a significant difference from the basic EDO.

coming larger and larger relative to the distance between the best solution and the optimal solution.

Adaptation for a Better Result.   With a fine tuning mechanism, such as the step size adaptation, this problem is eliminated (see Figure 7.1(b)). It can also be observed that the adaptive step size feature produces a population

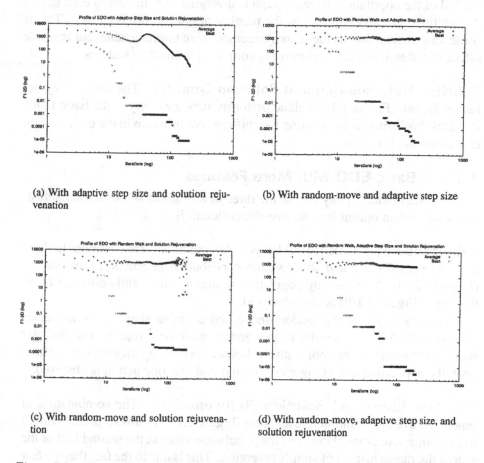

(a) With adaptive step size and solution reju-venation

(b) With random-move and adaptive step size

(c) With random-move and solution rejuvena-tion

(d) With random-move, adaptive step size, and solution rejuvenation

Figure 7.2. The relative performance advantages of step size adaptation, random-move, and solution rejuvenation strategies. (a) The combination of step size adaptation and solution rejuvenation does not produce better results than step size adaptation alone; (b) random-move and step size adaptation are comparable with step size adaptation alone, but the population average does not converge any more; (c) random-move and solution rejuvenation together not only produce the worst result, the population also becomes unstable; (d) all three strategies together gives a better solution than step size adaptation alone, and the population average does not diverge but converges very slowly.

that is converging at a faster rate than the basic EDO. This is further evidence of efficiency gained in utilizing the probability matrix.

Random Move and Population Diversity. The random-move be-havior (Figure 7.1(c)) seems to be making a more progressive improvement on the best solution than the basic EDO. A distinctive feature of random-move is that it produces a non-converging, if not diverging, population. Notice that the tail of the population average graph is diverging and the search ends before the set limit, at 159 iterations, as the number of entities drops to zero. This is somewhat undesirable as the search becomes more unpredictable, but this side effect vanishes when random-move is combined with other features.

Solution Rejuvenation and Solution Quality. The solution rejuvenation feature (Figure 7.1(d)) alone performs almost exactly as the basic EDO. It seems this feature is not making a significant contribution in the optimization of a unimodal function.

7.5.1 Basic EDO with More Features

The combination of any two of the three new features shows some interesting results when optimizing the two-dimensional f_1.

Adaptation and Random Move. Adaptive step size continues to show the best performance and shadows the contribution of solution rejuvenation (Figure 7.2(a)). A converging population similar to, but slightly different from, the one in Figure 7.1(b) is also observed.

The initial stage of the random-move and adaptive step size combination (Figure 7.2(b)) is very similar to the random walk only search. But the later stage of the search using both features shows a very steady improvement. However, the population is no longer converging and less fluctuation is observed.

Random Move and Solution Rejuvenation. The combination of random-move and solution rejuvenation (Figure 7.2(c)) shows a profile similar to random-move alone. However, the population value at the second half of the search fluctuates instead of simply diverging. This is due to the fact that the few good solutions are not eliminated from the system and allowed to reproduce beyond the time when active entities have become extinct.

Best Result and a Stable Population. When all three new features are used, EDO achieves the smallest value and the population shows a slight converging behavior (Figure 7.2(d)).

7.5.2 Clustering around the Optimal Solution

This subsection will show a phenomenon of entity clustering around optimal solutions. Specifically, it uses the two-dimensional version of f_1 and f_4. For the sake of illustration, Figures 7.3 and 7.4 present the shapes and planforms of f_1 and f_4, respectively.

Figures 7.5 and 7.6 show the distribution of entities as used to optimize the two-dimensional version of f_1. They show the behavior of the entities from a different perspective. Five entities are positioned randomly on the solution

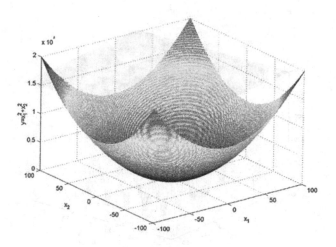

(a) The two-dimensional version of f_1

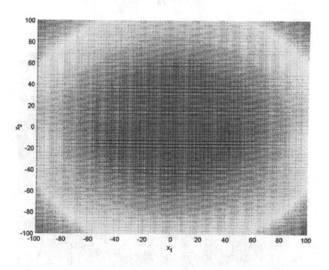

(b) The planform of the two-dimensional f_1

Figure 7.3. The two-dimensional unimodal function f_1.

space: three with x-value around -80, one around -20, and one around zero, but none is within the central 20 by 20 zone around the origin.

As the search progresses to iteration 10, more and more entities appear inside the central zone. The entities there will quickly reproduce, providing some positive feedback to their parents. This trend continues between iterations 10

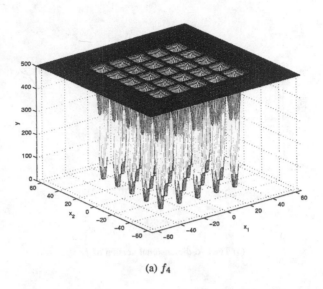

(a) f_4

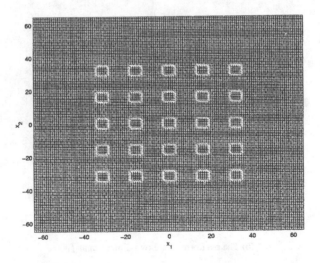

(b) The planform of f_4

Figure 7.4.   The multimodal function f_4.

and 80. Note that there are entities throughout the solution space. This is due to the diffusion operations of rational-move and random-move.

The diameter of the occupied area increases between iterations 10 and 30, and shrinks between iterations 30 and 80. This is the phase where the landscape

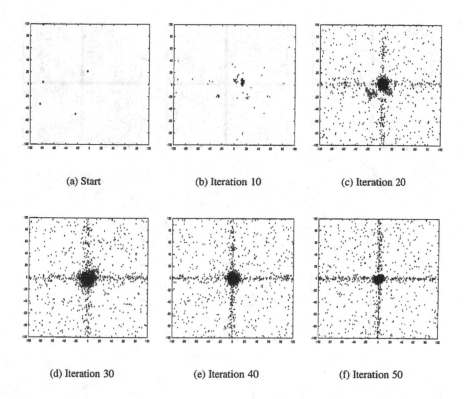

(a) Start   (b) Iteration 10   (c) Iteration 20

(d) Iteration 30   (e) Iteration 40   (f) Iteration 50

Figure 7.5. The distribution of entities in the search space for two-dimensional unimodal function f_1. The figures show a gradual increase in the concentration of entities around the optimal solution, which is the origin, and along the axes, which are the locally suboptimal solutions.

information has been learned. The entities have learned the direction of the fitness landscape in the first period. Then, they zoom in to the origin (the optimal solution) in the second period.

The third observation worth noting is the increase in the number of entities along the axes, which can be considered suboptimal solutions, as one of the two variables is in its lowest value.

The number of entities is at its highest point around iteration 80, but drops drastically in the next 10 iterations. This is because entities not around the origin or along the axes are less fit than the population average and they are eliminated eventually. As the entities continue to move towards the origin, the entities that are further away from the origin are eliminated, leaving only few entities inside the ± 0.01 range at iteration 220.

Figures 7.7 and 7.8 show the distribution of entities for f_4. EDO spends the initial 50 iterations to sample the search space, and builds up the probability

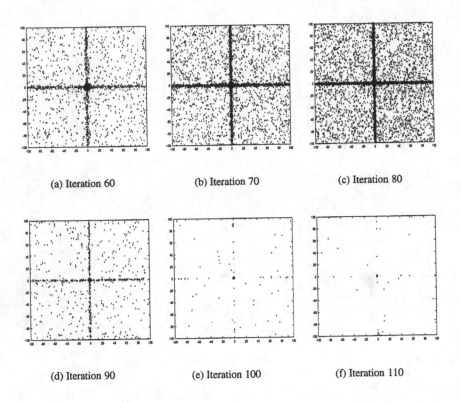

(a) Iteration 60   (b) Iteration 70   (c) Iteration 80

(d) Iteration 90   (e) Iteration 100   (f) Iteration 110

Figure 7.6. The distribution of entities in the search space for two-dimensional unimodal function f_1. As the search progresses towards the end, the number of entities remaining around the suboptimal solutions decreases.

matrix. At iteration 50, a rough impression of 25 clusters that correspond to the 25 local minima can be seen (Figure 7.7(d)). The number of entities around the local minima continues to grow until iteration 150 when the diameter of the clusters starts to diminish. Similar to the patterns in the unimodal case, the number of entities along the grid line grows as they are the local minima in the close-range neighborhoods. EDO stops at iteration 200 when the global minimum is found.

7.5.3   Summary

Adaptive step size achieves the best solution followed by random-move, if the function value is chosen as the evaluation criterion for the features. When more than one feature are considered, all three features together have a slight advantage over adaptive step size alone. It can be concluded that step size adaptation helps exploit existing solutions, while random-move behavior is useful for exploring new areas in the solution space. This is one of the most impor-

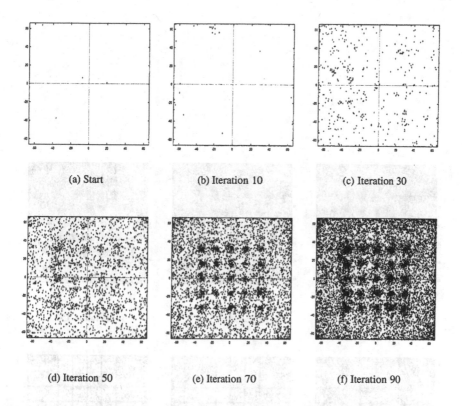

(a) Start   (b) Iteration 10   (c) Iteration 30

(d) Iteration 50   (e) Iteration 70   (f) Iteration 90

Figure 7.7. The distribution of entities in the search space for the multimodal function f_4. Similar to the situation with the unimodal function, there is a gradual increase in the concentration of entities around the suboptimal as well as the optimal solutions.

tant findings from this initial testing. Moreover, entities will tend to spread randomly across the search space. But once a suboptimal solution is found, they will be reluctant to leave there. The random-move behavior then helps push the entities out of the suboptimal region.

7.6. Experimentation

The EDO algorithm has been used to optimize the above four benchmark functions. Several forms of the functions, except function f_4 (which has only two free parameters), are tested by increasing the number of free parameters. The intention is to test the performance and scalability of EDO.

7.6.1 Unimodal Functions

EDO has been tested on six f_1 functions of 5 to 30 dimensions in increments of 5 dimensions. Figure 7.9(a) shows the development of the best solution over the iterations. In all the six experiments, EDO can successfully find a

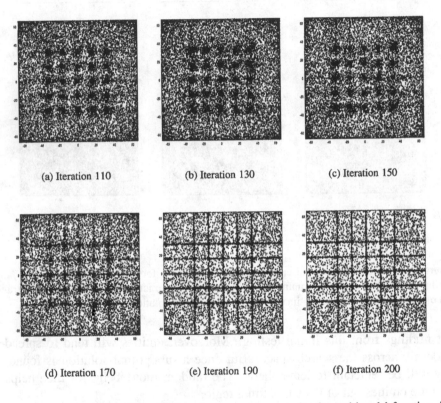

(a) Iteration 110   (b) Iteration 130   (c) Iteration 150

(d) Iteration 170   (e) Iteration 190   (f) Iteration 200

Figure 7.8.   The distribution of entities in the search space for the multimodal function f_4. The search stops at iteration 200 when the target is reached. If the search were to progress beyond iteration 200, the number of entities, especially around the suboptimal solutions, would decrease gradually.

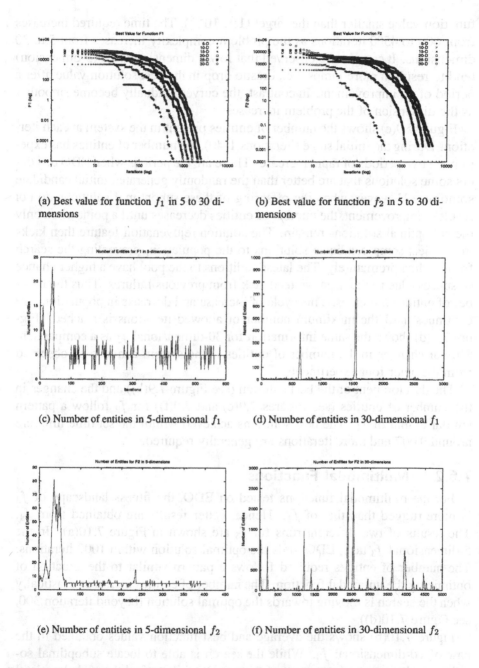

(a) Best value for function f_1 in 5 to 30 dimensions

(b) Best value for function f_2 in 5 to 30 dimensions

(c) Number of entities in 5-dimensional f_1

(d) Number of entities in 30-dimensional f_1

(e) Number of entities in 5-dimensional f_2

(f) Number of entities in 30-dimensional f_2

Figure 7.9. The best function value found by EDO and the number of entities used at each iteration for two unimodal functions, $f_1(x)$ and $f_2(x)$, in 5 to 30 dimensions. (a)-(b) The number of iterations required to reach the target increases with the dimension of the problem; (c)-(f) The number of entities required over iterations follows a cycle. A drastic increase corresponds to the time when a suboptimal solution is located and a decrease corresponds to a period of fine-tuning the probability matrix.

function value smaller than the target (1×10^{-6}). The time required increases from 600 to 4500 iterations as the problem complexity increases from 5 to 30 dimensions. It can also be observed that lower dimension tasks (5-dimension) tend to result in more jumps, i.e., drastic drop in the best solution value after a period of no improvement. In contrast, the curves gradually become smoother as the dimension of the problem increases.

Figure 7.9(c) shows the number of entities present in the system at each iteration. During the initial stage (iterations 1-40), the number of entities has experienced two periods of rapid increase. These are the periods when EDO discovers some solutions that are better than the randomly generated initial candidate solutions. However, when the offspring of these suboptimal solutions do not result in improvement, the number of entities decreases until a point when only the suboptimal solutions remain. The solution rejuvenation feature then kicks in to inject some candidate solutions to the population, preventing the search from ending prematurely. The latest additions to the pool have a higher chance of success due to the negative feedback from previous failures. Thus the number of entities increases. This cycle of increase and decrease in population size continues until the maximum number of allowed iterations is reached. Figure 7.9(d) shows the same information for 30-dimensional f_1 as a comparison. Similar changes in the number of entities in the system can also be observed from the other four experiments.

The development of the best solution (see Figure 7.9(b)) and the changes in the number of entities (see Figures 7.9(e) and 7.9(f)) for f_2 follow a pattern similar to that of f_1. The best solutions achieved within the set time limit are around 0.002 and more iterations are generally required.

7.6.2   Multimodal Functions

For the multimodal functions tested on EDO, the fitness landscape of f_3 is more rugged than that of f_4. Hence, better results are obtained from f_4. The results of two different runs for f_3 are shown in Figure 7.10(a). In the 5-dimensional f_3 task, EDO finds the optimal solution within 1000 iterations. The number of entities required follows a pattern similar to the situation of optimizing the unimodal function: The number of entities increases drastically when the search is moving towards the optimal solution (beyond iteration 300, see Figure 7.10(d)).

Figure 7.10(b) shows the average and best function values obtained in the case of 10-dimensional f_3. While the search is able to locate suboptimal solutions and to escape from most of them, EDO fails to find the global optimal within the given time limit. Note that the population average fluctuates much more rapidly and sometimes even becomes very close to the best population. As the search progresses without making much improvement, the number of entities will decrease due to, for example, the expiry of lifespan. In the worst

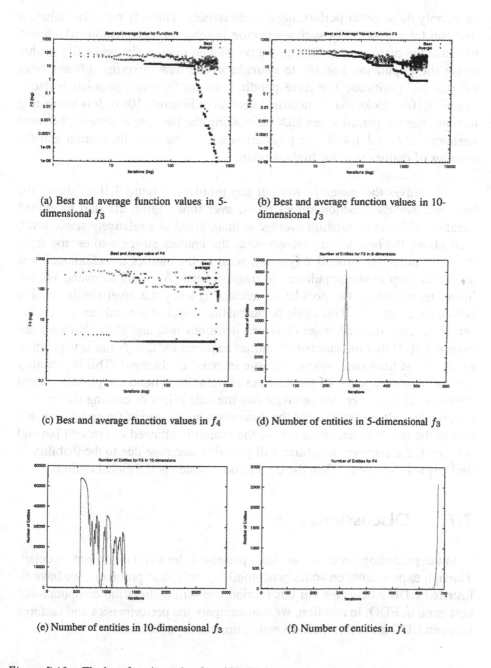

(a) Best and average function values in 5-dimensional f_3

(b) Best and average function values in 10-dimensional f_3

(c) Best and average function values in f_4

(d) Number of entities in 5-dimensional f_3

(e) Number of entities in 10-dimensional f_3

(f) Number of entities in f_4

Figure 7.10. The best function value found by EDO and the number of entities used at each iteration for two multimodal functions, $f_3(x)$ and $f_4(x)$. (a)-(c) The population tends to diverge at the end of the search in f_3, showing that EDO attempts to improve beyond a suboptimal solution; (d)-(f) The total number of entities in a generation increases with the number of successful entities, but will drop as some of them are not making sufficient progress.

case, only those better performing parents remain. The only primitive behavior that can take an effect in such a situation is solution rejuvenation, which will increase the number of entities in the population once again, and will possibly cause the population average to increase as the new offspring diffuses from the parents' positions. The corresponding plot for the entity population size in Figure 7.10(e) looks like a magnified version of Figure 7.10(d). It is interesting to note that the period when EDO is making the best improvement (between iterations 500 and 1000), the population shows the least fluctuation and the number of entities is at the highest point.

EDO solves the easier f_4 without any problem. Figure 7.10(c) shows the best and average function values found over time. From the start to around iteration 100, the population average is maintained at a relatively stable level, well above the best value. However, as the entities struggle to escape from the local minimum around 2 (y-axis), most of the entities are eliminated and a sudden drop in the population average occurs. As EDO recovers, via solution rejuvenation, the population average reaches the level similar to that before iteration 100. This cycle of fluctuation happens several times until the search reaches its final stage (between iterations 600 and 700). Notice from Figure 7.10(f) that the number of entities required for this f_4 run is very small except at the final stage where a drastic increase is observed. This is probably due to the convergence of most, if not all, entities around the global optimal solution. This convergent behavior has the side effect of causing the entities to reproduce. It is obvious that the downward going side of the spike is absent due to the reach of the set target. If the search is allowed to proceed beyond this point, the number of entities will probably decrease due to the inability to find any solution better than the current best, which is the global optimum.

7.7.   Discussions

In the preceding sections, we have presented the EDO algorithm in detail. Through experiments on some benchmark optimization problems, we have illustrated EDO's behavior. In this section, we will address the computational cost issue in EDO. In addition, we will compare the performances and features between EDO and other well-known optimization algorithms.

7.7.1   Computational Cost

In what follows, we will examine the space and time complexity of the EDO algorithm.

7.7.1.1 Space Complexity

Let $P^{(t)}$ denotes the population of entities at iteration t, which consists of the active subpopulation, $P_a^{(t)}$, and the inactive subpopulation, $P_i^{(t)}$. Therefore,

$$|P^{(t)}| = |P_a^{(t)}| + |P_i^{(t)}|, \qquad (7.28)$$

where $|\cdot|$ denotes the size of the population concerned. The size of the two subpopulations changes due to the various local behavioral responses of the entities in the previous iteration. For the inactive subpopulation, the number of entities is given by:

$$|P_i^{(t)}| = |P_i^{(t-1)}| \times (1 - \nu_i) + \mu |P_a^{(t-1)}|, \qquad (7.29)$$

where $0 \le \nu_i \le 1$ and $0 \le \mu \le 1$ are the percentage of inactive entities at iteration $(t-1)$ that no longer have active offspring and whose fitness is below average, and the percentage of successful active entities at iteration $(t-1)$ that will reproduce and become inactive, respectively.

For the active population, all successful entities will replicate themselves several times and become inactive, while some below average or old ones are eliminated. The number of active entities at iteration t is given by:

$$
\begin{aligned}
|P_a^{(t)}| &= |P_a^{(t-1)}| + \mu |P_a^{(t-1)}| \times (Q(x) - 1) - \nu_a |P_a^{(t-1)}| \\
&= |P_a^{(t-1)}| \times (1 + \mu(Q(x) - 1) - \nu_a), \qquad (7.30)
\end{aligned}
$$

where $0 \le \nu_a \le 1$ is the percentage of active entities at iteration $(t-1)$ that are either too old or under-performing, μ is the percentage of good performers that have reproduced and become inactive, and $Q(x)$ is the function that gives the number of replicas an active entity can produce.

The upper bound for the size of the inactive population occurs when no inactive entity in the previous iteration is eliminated and all active entities will reproduce, i.e., $\mu = 1$ and $\nu_1 = 0$. Hence,

$$\lceil |P_i^{(t)}| \rceil = |P_i^{(t-1)}| + |P_a^{(t-1)}|. \qquad (7.31)$$

Similarly, the upper bound for the size of the active population is reached when no active entities is eliminated because of its age or fitness, and all active entities become parents and reproduce the maximum number of times, i.e., $\mu = 1$ and $Q(x) = \Omega$. As a result,

$$\lceil |P_a^{(t)}| \rceil = |P_a^{(t-1)}| \times \Omega. \qquad (7.32)$$

In other words,

$$\lceil |P^{(t)}| \rceil = |P_i^{(t-1)}| + |P_a^{(t-1)}| \times (1 + \Omega)$$

Algorithm 7.2 The adaptive simulated annealing (ASA) algorithm.

Initialize population $P^{(t)}$;
Initialize temperature $u^{(t)}$ according to the annealing schedule;
while ($u^{(t)}$ > TerminationTemp) and (a target error margin is not reached)
do

  Evaluate population $P^{(t)}$;
  Generate new population $P^{(t)}$;
  Reanneal temperature $u^{(t)}$ according to the annealing schedule;
end while

$$
\begin{aligned}
&= (|P_i^{(t-2)}| + |P_a^{(t-2)}|) + |P_a^{(t-2)}| \times \Omega \times (1 + \Omega) \\
&= |P_i^{(0)}| + |P_a^{(0)}| \times (1 + \Omega + \Omega^2 + \ldots + \Omega^t) \\
&= |P_a^{(0)}| \times \frac{(\Omega^{t+1} - 1)}{(\Omega - 1)}, \quad (\text{since} |P_i^{(0)}| = 0).
\end{aligned} \tag{7.33}
$$

Therefore, the maximum offspring parameter, Ω, is the most important factor that will affect the overall space requirement in EDO.

7.7.1.2   Time Complexity

The number of function evaluations is the same as the total number of active entities throughout the search. Therefore, the total number of evaluations performed by EDO during t iterations is as follows:

$$
\text{Total Evaluations} = \sum_{j=0}^{t} |P_a^{(j)}| = |P_a^{(0)}| \times \frac{(\Omega^{t+1} - 1)}{\Omega - 1}. \tag{7.34}
$$

7.7.2   Feature Comparisons

Several well-known search algorithms share some common features with EDO. Without going into too much detail of the algorithms, this section attempts to highlight some of their major differences. Interested readers are encouraged to study the respective references. Pseudocodes for the search algorithms are included in the following discussion in a format similar to the pseudocode for EDO (see Algorithm 7.1) for easy reference.

7.7.2.1   Simulated Annealing

Adaptive simulated annealing [Ingber, 1996] is similar to the classic simulated annealing. The major enhancements are in the reannealing process where the annealing and the control temperature are adjusted (see Algorithm 7.2).

Apart from the difference in population size, both ASA and EDO maintain a schedule for modifying the systems parameters. In ASA, the temperature of the cooling process is predetermined to go down gradually. The counterpart in EDO is the modification of the step size. However, EDO's step size can go up as well as down depending on the progress over a period of time.

Another major difference between ASA and EDO lies in the way they accept a worse candidate solution as the starting point of the next iteration. While ASA will accept this candidate based on certain probability, EDO will always accept it. The aging mechanism in EDO allows the system to get rid of poor performers after they are consistently performing badly over a period of time.

7.7.2.2 Evolutionary Algorithms

A genetic algorithm (GA) has a population of chromosomes (usually of fixed size) that are used to represent a candidate solution to the problem at hand [Holland, 1992]. The chromosome can be imagined as a compartmentalized representation of the parameters for the current problem domain. Each randomly generated chromosome in the initial population is given a fitness value, which is an assessment of its goodness-of-fit to the problem by a fitness function. The fitness function collectively represents the goal to be achieved by GA. In each cycle, or generation, of a GA run, pairs of chromosomes are selected based on the fitness for recombination (crossover) and mutation to create new chromosomes. Recombination involves splicing up two parents and exchange part(s) of their chromosomes. Mutation involves changing some part of a chromosome to a value allowable for the chosen gene(s). The processes of recombination and mutation are performed based on some probabilities (usually remaining constant throughout the run). Newly created chromosomes are assessed using the fitness function, and will replace their parents if they have higher fitness values. The simulated evolution cycle repeats until a predefined number of generations have passed or certain criteria are met.

There are many strands in the class of evolutionary algorithms, which are mostly population-based. However, most of them maintain a fairly common framework based on the survival of the fittest principle (see Algorithm 7.3). They differ in the choice of the so called genetic operators and in the implementation of the algorithms. A common departure from the usual practice is to use mutation as the only genetic operator (see Algorithm 7.4). EDO is different from EA in the following ways:

1. The population size of EA is usually fixed, while EDO maintains a variable-size population. Such a method allows the weaker entities that survive to the next iteration to wander to a better position, and hence they are free from being trapped by local suboptimal solutions.

2. A variable mutation rate, realized by the diffusion step size, allows EDO to migrate smoothly between exploration and exploitation. Moreover, most EAs give a high priority to making small changes rather than large changes.

Algorithm 7.3 A typical population-based evolutionary algorithm in which offspring entities are created through the recombination of parents.

$t = 0$;
Initialize population $P^{(t)}$;
Evaluate population $P^{(t)}$;
while the termination condition is not met **do**
    $t++$;
    Select parents $\bar{P}^{(t)}$ from population $P^{(t)}$;
    Recombine parents $\bar{P}^{(t)}$;
    Mutate parents $\bar{P}^{(t)}$;
    Evaluate parents $\bar{P}^{(t)}$;
    Generate new population $P^{(t)}$ based on parents $\bar{P}^{(t)}$ and old population $P^{(t)}$;
end while

Algorithm 7.4 A typical mutation only evolutionary algorithm, such as evolutionary programming and evolution strategies. It should be noted that parameters controlling the mutation process is also encoded as some of the objects to be optimized by the evolutionary algorithm. As a result, self-adaptive behavior is obtained.

$t = 0$;
Initialize population $P^{(t)}$;
Evaluate population $P^{(t)}$;
while the termination condition is not met **do**
    $t++$;
    Generate $\bar{P}^{(t)}$ through mutating population $P^{(t)}$;
    Evaluate $\bar{P}^{(t)}$;
    Generate new population $P^{(t)}$ based on $\bar{P}^{(t)}$ and old population $P^{(t)}$;
end while

The latter may be able to use fewer steps for searching if larger steps are possible.

3. Selection in EA is a population-wide operation where all entities in the population are sorted according to their fitness. However, selection in EDO is entirely local, which compares only the performances of an entity and its parent. This local selection process is computationally less costly and offers a better chance for parallel implementation.

4. The probability matrix in EDO allows entities to capture the trend of the fitness landscape, and points a further search towards a potentially fruitful direction. This information is again local to an entity and shared only by

Algorithm 7.5 The particle swarm optimization (PSO) algorithm, which is similar to an evolutionary algorithm. The major difference is the way mutation is performed. While no control parameter is being optimized, the best solution so far is used as a reference point during the generation of the next candidate solution.

$t = 0$;
Initialize population $P^{(t)}$;
while the termination condition is not met **do**
    Evaluate population $P^{(t)}$;
    Select the best particle(s) $b^{(t)}$ from population $P^{(t)}$;
    Based on $P^{(t)}$ and $b^{(t)}$, generate new population $P^{(t)}$ through mutations;
    $t + +$;
end while

an entity's offspring. In contrast, any trend in EA can be captured only by observing all entities in the population.

5. The chance to reproduce and remain in the population decreases with the entity's fitness in the case of EA. In contrast, the notion of success is limited to the local sense in EDO. This allows a high diversity in the population. Random-move plays a stronger role in EDO than in EA, as an instrument to escape from suboptimal solutions.

7.7.2.3 Particle Swarm Optimization

Particle swarm optimization (PSO) is similar to EDO in that individual particles make their own local decisions without consulting any other particles except one [Kennedy, 1997]. In more detail, individual particles require two pieces of information to make their decision regarding the next position to assume: the velocity (direction and speed) of the last move and that of the overall best particle. Therefore, some kind of global information is always required (see Algorithm 7.5). Although communications among entities of the same lineage are required in EDO, they do not depend on any global information.

7.7.2.4 Ant Colony Optimization

Ant colony optimization (ACO) is an optimization algorithm based on the principle of autocatalysis as learned from ants [Dorigo et al., 1996]. A food foraging ant will lay a pheromone trail on its way home from a food source. Similarly, individual entities in ACO, when solving combinatorial optimization problems share information by building up the relative importance of individual path segments incrementally (see Algorithm 7.6). Landscape trend information sharing in EDO is explicit and local. This is achieved via the probability matrix, which is used and updated by all offspring of an entity.

Algorithm 7.6 The ant colony optimization (ACO) algorithm. The special feature of ACO is the presence of self-organized information (pheromone trail) shared among ants.

$t = 0$;
Initialize population $P^{(t)}$;
Initialize trail $R^{(t)}$;
while the termination condition is not met **do**
    Evaluate population $P^{(t)}$;
    Update trail $R^{(t)}$;
    According to trail $R^{(t)}$, construct new population $P^{(t)}$;
    Based on $P^{(t)}$ and local search, generate new population $P^{(t)}$;
    $t + +$;
end while

Algorithm 7.7 The cultural algorithm. A more elaborate information sharing scheme via belief updating is involved.

$t = 0$;
Initialize population $P^{(t)}$;
Initialize belief $B^{(t)}$;
while the termination condition is not met **do**
    Evaluate population $P^{(t)}$;
    According to belief $B^{(t)}$, vote on population $P^{(t)}$;
    Adjust belief $B^{(t)}$;
    Evolve population $P^{(t)}$, while affecting belief $B^{(t)}$;
    $t + +$;
    Select new population $P^{(t)}$ from old population $P^{(t-1)}$;
end while

7.7.2.5   Cultural Algorithms

Human beings pass experiences and knowledge to the next generation and beyond by means of culture [Reynolds, 1994]. This form of knowledge sharing has been captured in the cultural algorithm (see Algorithm 7.7). EDO also takes the view that knowledge gained by one entity is useful to others. As such, the direction of change is passed to others via the probability matrix, which is consumed by entities within the family. This is particularly useful in a population-based search algorithm, where many sites in the fitness landscape are being explored in parallel, as the direction information in one site is usually irrelevant to another site.

7.8.   Summary

This chapter has presented an AOC-based method, called evolutionary diffusion optimization (EDO), for global optimization. From the previous chapters, we have seen that an AOC-based computational system usually involves a large number of distributed autonomous entities. Through local and nonlinear interactions among entities, the behavioral results of entities will be self-aggregated and consequently certain complex patterns or behaviors emerge. EDO utilizes this mechanism and relates the resulting emergent complex patterns or behaviors to the solutions to an optimization problem at hand. Hence, EDO is well suited for solving the type of optimization problems that are characterized as being large-scale, highly distributed.

7.8.1   Remarks on EDO

In EDO, each entity is equipped with the primitive behaviors to diffuse to its local neighborhood and reproduce. In addition, it can also choose to perform random-move behaviors. Decisions on the course of behaviors are made by an entity based on the common information shared with its parent and its siblings. The common information contains the likelihood estimates of finding a good solution in a certain direction, and is updated by every member of the family – reinforcing positively the good moves and negatively the bad moves. EDO also has a mechanism to adapt its step size during the search.

The analysis of EDO performance shows that it can maintain a high diversity in the population of entities throughout the search – a crucial feature to avoid premature convergence.

Our experiments reveal that EDO is able to automatically maintain a good balance between exploration and exploitation. This is made possible in EDO through probabilistically performing random-move behaviors that help maintain the population diversity. At the same time, the ability to automatically adapt the search step size has been proven to be very useful.

7.8.2   Remarks on AOC by Self-Discovery

AOC-by-self-discovery is the same as AOC-by-prototyping except that the process of trial-and-error in AOC-by-self-discovery is automated (see Figure 7.11). The automation of the prototyping process is achieved by having one autonomous entity to control or fine-tune the parameters of other autonomous entities. The EDO example described in this chapter shows that AOC-by-self-discovery is a viable approach. The steps for engineering this kind of AOC algorithm are the same as those in Section 6.6.2 with the addition of one rule, that is, systems parameters are self-adapted according to some performance feedback.

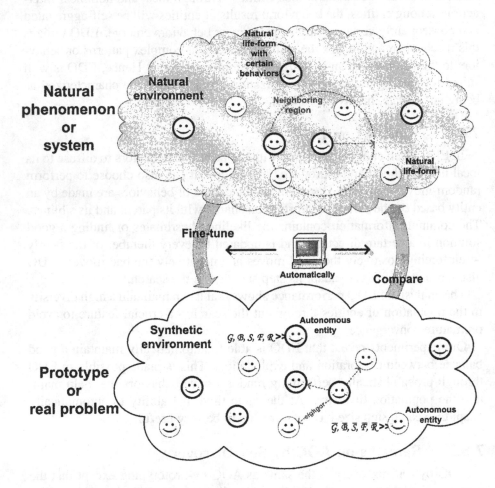

Figure 7.11. The AOC-by-self-discovery approach. As compared to AOC-by-prototyping (see Figure 6.34), here the trial-and-error process, i.e., repeated *fine-tune* and *compare* steps, is automatically performed by the system (as symbolized by a computer sign in the figure).

Exercises

7.1 Identify the following elements in EDO:

(a) Stochastic behavior;

(b) Shared knowledge;

(c) Inter-entity communication;

(d) Self-discovery.

How would you improve them?

7.2 List the major similarities and differences between EDO and the algorithms mentioned in Section 7.7.2.

7.3 Population-based search algorithms often face the risk of premature convergence. Is EDO facing the same threat? Does EDO have any mechanism to avoid premature convergence? Can that mechanism be improved?

7.4 Growth in the population size is a potential problem in EDO.

(a) Suggest ways to control the size globally;

(b) Design a method to control the population within a proximity of a local optimal solution.

7.5 As the global information in EDO, step size helps control how exploration is done. Is there any benefit in making the step size information local to an entity, or to a family tree?

7.6 Fitness is a key measurement to direct a search algorithm. But in some situations, such as the blind 0-1 knapsack problem [Goldberg and Smith, 1987], the fitness function is non-stationary. Modify EDO to handle this new requirement.

7.7 Modify the formulation of EDO to tackle combinatorial optimization tasks, such as the traveling salesman problems [TSP, 2002].

7.8 Multi-objective optimization problems [Eschenauer et al., 1986] deal with a set of optimization criteria instead of just one. Mathematically, $F(x)$ in Equation 7.7 becomes:

$$F(x) = [F_1(x), F_2(x), F_2(x), \cdots, F_m(x)]. \qquad (7.35)$$

Modify EDO to handle the m competing criteria. Read [Coello, 2002] and the references within it to get an idea on some GA-like methods.

Chapter 8

Challenges and Opportunities

Technological advances have given birth to the Internet age, which has revolutionized the ways in which people interact and companies conduct their business. It also helps open our doors to new issues in large-scale scientific or social computing applications and how they can be developed. Just imagine the amount of data that a Mars exploratory robot or an Internet information spider needs to process.

The availability of distributed computing resources adds new power to problem solvers. The time is ripe to call for a new paradigm shift in ways of solving modern, complex problems.

Drawing on the notion of autonomy oriented computing (AOC) as introduced in [Liu, 2001, Liu and Tsui, 2001], this book has laid down both theoretical and engineering foundations for building highly distributed yet effective artificial complex systems.

8.1. Lessons Learned

We have provided, in Chapters 2 and 3, a general overview of the AOC modeling methodology. We have seen, in Part II of this book, some representative case studies on how to implement the general approaches to AOC. It is instructive to point out that the indiscernible ingredients of AOC problem solving include:

- A population of autonomous entities;

- A behavior model for the autonomous entities;

- A model of local interactions between entities and their environment;

- A definition of roles and responsibilities of the environment;

- A set of criteria for measuring and self-organizing the nonlinear behavior of AOC problem solving.

With the models and definitions in place, what remains is to let the system of autonomous entities to run and interact according to the models and observe its emergent behavior. It is worth underlining the main lessons learned from the cases studies described in Part II.

Autonomous behavior is local. The n-queen example described in Chapter 2 shows that primitive behaviors based on some limited amount of information observable from the limited range surroundings are sufficient for solving a global constraint problem. The implication is that we can save a lot of resources normally required by problem solvers adopting a centralized planner.

Primitive behavior is neighbor-driven. Similarly, the image segmentation problem in Chapter 2 shows that controlling the direction of entity propagation can help nonlinearly amplify a desirable pixel labeling behavior and produce more entities of this locally successful entity type. The behavior of an autonomous entity does not need to be driven by some global information in order to produce a desirable global behavior.

Autonomous behavior is self-directed. The Web foraging entities described in Chapter 6 have two measures associated with them: motivational support and interest profile. The values of these two measures directly affect how an entity picks its next move.

Stochastic behavior is beneficial. All the AOC algorithms described in this book share a common feature – the presence of stochastic behavior. For example, the ERE model (Chapter 5) has a random-move behavior, which randomly chooses a new position (for n-queen problems) or a new variable value (for SAT problems). Similarly, the entity in EDO (Chapter 7) has a choice to select a random-move behavior when it has not been making progress for some time.

Stochastic behaviors enable an autonomous entity to get out of local minima. As a result, an AOC algorithm will have a better chance of locating a potentially better solution in the search space.

Some behaviors can be self-reinforcing. Chapter 7 has illustrated the implementation of an AOC-by-self-discovery method. As discussed previously, a search algorithm must find an appropriate step size in order to perform the search efficiently. However, a one-for-all step size is bound to err as the search landscape varies drastically among different sites.

The step size adaptation behavior is, therefore, implemented and it considers the degree of success in locating a locally better solution as an indication of whether a bigger or smaller step size is required. By including such a behavior adaptation strategy, one can take a lot of guess work away from the trial-and-error process that is normally required in an AOC-by-prototyping method.

A means of separating the good from the poor solutions is required. Population explosion is probably one of the most troublesome problems in an AOC algorithm. While better solutions are always welcome and seen as stepping stones to an even better solution, it is not advisable to keep replicating them indiscriminately. EDO in Chapter 7 has implemented a step function to grant different numbers of copies that entities with good performance can replicate themselves. On the other hand, those entities with poor performance are eliminated gradually. The image segmentation example in Chapter 2 and EDO in Chapter 7 associate an age to every entity, and persistently unsuccessful entities will eventually reach their natural end.

Exchange of information is important but can be minimal. While information sharing is crucial to the success of the EDO algorithm, the amount of information flow can be minimal. In the presence of spatially distributed computing resources, this feature becomes very important.

Autonomous entities in EDO limit their communications indirectly to those that share the same starting point (or parent). While the parent uses all the information fed by its offspring entities to bias its choice of behavior, the entities communicate with their respective parent to obtain the current best solution. In other words, all the autonomous entities in the search space are affecting each other indirectly.

Self-organization is the key to success. The centerpiece of an AOC algorithm is the notion of self-organization. When autonomous entities with self-directed behaviors are allowed to aggregate and react to the information and stimulation of other autonomous entities, a desired global behavior emerges as a result.

For example, the Web foraging regularity modeling example presented in Chapter 6 models Web content distribution, the internal 'state of mind' of entities, and their behavior, and is able to yield the scale-free behaviors that match observations in real life. The success in building a model with such an emergent behavior helps researchers explain Web surfing behavior. Similar complex systems modeling methods can be adopted in other studies, such as stock investor behavior or car driving behavior analysis.

The above notes have highlighted some of the key elements to be built into an AOC system. They are certainly useful and have been demonstrated in the previous chapters. Several areas remain to be extended in order to fully exploit the benefits of AOC. The remainder of this chapter sketches some directions, both theoretical and practical, for future research.

8.2.   Theoretical Challenges

New approaches to systems dynamics and performance measurements are particularly needed so that clearer guidelines can be developed to help practitioners gain better insights into AOC, and AOC-by-self-discovery in particular. The measurements of emergence, evolvability, self-organization, tractability, and scalability in AOC are useful for tracking the progress of AOC. Theories on the formation of roles and social structures in a community of autonomous entities would expand the capability of an AOC system.

The strength and weakness of AOC need to be assessed formally by comparing with other multi-entity paradigms to establish clear insights into the benefits of AOC. Benchmark AOC problems should be identified for this purpose.

8.3.   Practical Challenges

To foster and encourage the adoption of AOC for problem solving and complex systems modeling, more real-world applications as well as the characterization of potential areas need to be identified.

More guidelines and tools for developing AOC are needed so that people can readily benefit from this new computing paradigm. The requirements for simulation environments, languages, and data structures in AOC need to be studied so that a more efficient implementation of AOC can result. Other implementation issues that need to be addressed include: architecture, visualization of activities, and the design of local and global nonlinear interaction rules.

A few decisions that need to be made when implementing an autonomy oriented computing application. They include: hardware and software environments, update schedule, management services, and visualization. This section discusses the relevancy of these issues to the AOC approaches.

8.3.1   Hardware and Software Environments

AOC usually involves more than one component in the system. These components include autonomous entities and static or dynamical environments. Implementing an AOC system in a single processor machine requires the support of virtual parallelism. Modern operating systems and programming languages support multi-thread technique. This allows slicing up CPU cycles and allocates them to the individual processes that represent components in the system. When multiple processors are available, individual components can be allocated to different processors on the same machine or across the network of processors with the support of some facilities, such as parallel virtual machine (PVM) [Geist et al., 1994, PVM, 1989] and message passing interface (MPI) [Snir et al., 1996, MPI, 1996]. However, it requires a central control program to coordinate task allocation and result consolidation. This makes parallel implementation of, for example, genetic algorithms possible without requiring expensive parallel machines.

With the popularity of peer-to-peer and grid computing networks, mobile entities can be sent to run on any machines over the Internet, provided that permission is granted by the host. Running simulated evolution in this way has been attempted [Smith and Taylor, 1998] and is a good starting point for those pursuing this line of research.

8.3.2 Update Schedule

An individual entity changes its state at each step based on its current state and its neighboring environment. In a synchronous update scenario, the current state of all individuals is frozen to allow all individuals to obtain state information and change states, if appropriate. The current state of the whole system is then updated and the system clock ticks, marking the beginning of the next step. In a parallel implementation, synchronization may become an overhead too big to handle. Alternatively, asynchronous updates that allow processes on each processor to proceed independently may be implemented. However, the choice of an update schedule and the choice of a hardware platform are related. If a multi-process hardware environment is chosen, synchronous updates would slow the simulation down as all processes have to start and stop at the same time.

8.3.3 Management Services

With the vast number of autonomous entities in an AOC system, AOC needs to keep track of the creation and deletion of objects. Moreover, a messaging mechanism is needed to facilitate message passing between objects. A central clock is also required to help the autonomous entities manage their state updates, no matter if it is synchronous or asynchronous. Some centralized whiteboards may also be needed if the autonomous entities are to share information in an implicit way and to contain a global view of the system's status. The whiteboards may also be used to simulate a dynamical environment in such systems as the ant system [Dorigo et al., 1996].

8.3.4 Visualization

Visualization is a good way for people running simulations to 'see' what is going on with the experiment. Items of interest that are related to individual autonomous entities include actual movements, states, actions taken, fitness, ages, etc. On the other hand, some global information (such as population size, best fitness, and average fitness) and some progress measurements (such as measurements of emergence, evolvability, diversity, and convergence) are also of interest to modelers. The visual display of such information will be of tremendous help to modelers to obtain a quick view of the system.

8.4.   Summary

We have presented in this book a new computing paradigm, called autonomy oriented computing (AOC), based on the notions of autonomy and self-organization in entities. AOC is intended to meet the demands of real-world computing that is naturally embodied in large-scale, highly distributed, locally interacting entities, as in sensor networks, grid computing, and amorphous computing. Nevertheless, as we have demonstrated through examples, AOC is also applicable to tackling conventional computing problems.

In this book, we have highlighted the most fundamental issues central to the design, formulation, and implementation of an AOC system. To recap, AOC has three general approaches with different objectives:

1. AOC-by-fabrication is similar to construction with a blueprint, where some more or less known complex systems phenomena are abstracted and replicated in problem solving or system modeling.

2. AOC-by-prototyping represents a trial-and-error approach to finding explanations to some complex behavior observations via autonomy oriented systems prototyping. Human involvement is sometimes intensive to fine-tune systems parameters.

3. AOC-by-self-discovery, on the other hand, is an autonomous problem solving approach that can fine-tune its own settings to suit the problem at hand. It requires less human intervention than AOC-by-prototyping, but represents the highest degree of uncertainty, as it is difficult to predict when the system will stop.

The advantages of using AOC to solve hard computational problems or to model complex systems behavior are apparent as we have mentioned at the beginning of this book (i.e., in Preface). By providing detailed methodologies and case studies, we hope to lay down the foundations for future AOC development, and at the same time, to stimulate general interests in this newly emerged, exciting computing field.

Exercises

8.1 Based on your understanding of AOC, review and summarize the key steps of using AOC in problem solving or complex systems modeling.

8.2 Based on your understanding of AOC, identify and explain:

(a) What are the advantages and disadvantages of AOC?

(b) What problem domains is AOC specially suitable for? What domains is it not?

8.3 From an engineering point of view, think and design a dedicated platform or programming language for AOC.

References

[Adamic, 1999] Adamic, L. A. (1999). The small world Web. In *Proceedings of the Third European Conference on Research and Advanced Technology for Digital Libraries (ECDL'99)*, volume 1696 of *LNCS*, pages 443–452. Springer.

[Adamic and Huberman, 1999] Adamic, L. A. and Huberman, B. A. (1999). Technical comment to 'Emergence of scaling in random networks'. *Science*, 286:509–512.

[Adamic and Huberman, 2000] Adamic, L. A. and Huberman, B. A. (2000). The nature of markets in the World Wide Web. *Quarterly Journal of Electronic Commerce*, 1:5–12.

[Adar and Huberman, 2000] Adar, E. and Huberman, B. A. (2000). The economics of surfing. http://www.hpl.hp.com/research/idl/abstracts/ECommerce/econsurf.html.

[Albert et al., 1999] Albert, R., Jeong, H., and Barabasi, A. L. (1999). Diameter of World Wide Web. *Nature*, 410:130–131.

[Angeline, 1995] Angeline, P. J. (1995). Adaptive and self-adaptive evolutionary computations. In Palaniswami, M., Attikiouzel, Y., Marks, R., Fogel, D., and Fukuda, T., editors, *Computation Intelligence: A Dynamic Systems Perspective*, pages 152–163. IEEE Press.

[Angeline, 1996] Angeline, P. J. (1996). Two self-adaptive crossover operators for genetic programming. In Angeline, P. J. and Kinnear, Jr. , K. E., editors, *Advances in Genetic Programming*, pages 89–110. MIT Press.

[Arlitt and Williamson, 1996] Arlitt, M. F. and Williamson, C. L. (1996). Web server workload characterization: The search for invariants. In Gaither, B. D., editor, *Proceedings of the ACM SIGMETRICS'96 Conference on Measurement and Modeling of Computer Systems*, pages 126–137. ACM Press.

[Ashby, 1966] Ashby, W. R. (1966). *Design for a Brain*. Chapman & Hall, 2nd edition edition.

[Ashby, 1947] Ashby, W. Ross (1947). Principles of the self-organizing dynamic system. *Journal of General Psychology*, 37:125–128.

[Bäck, 1992] Bäck, T. (1992). The interaction of mutation rate, selection, and self adaptation within a genetic algorithm. In Manner, R. and Manderick, B., editors, *Parallel Problem Solving From Nature*, pages 85–99. North-Holland.

[Bäck, 1994] Bäck, T. (1994). Parallel optimization of evolutionary algorithms. In Davidor, Y., Schwefel, H. P., and Männer, R., editors, *PPSN III: Proceedings of the Third Conference on Parallel Problem Solving from Nature*, volume 866 of *LNCS*, pages 418–427. Springer.

[Bäck, 1997] Bäck, T. (1997). Self-adaptation. In Bäck, T., Fogel, D. B., and Michalewicz, Z., editors, *Handbook of Evolutionary Computation*, chapter 7.1, pages 1–15. Oxford University Press.

[Bäck et al., 1997] Bäck, T., Fogel, D. B., and Michalewicz, Z., editors (1997). *Handbook of Evolutionary Computation*. Oxford University Press.

[Bäck and Schutz, 1996] Bäck, T. and Schutz, M. (1996). Intelligent mutation rate control in canonical genetic algorithms. In Raś, Z. W. and Michalewicz, M., editors, *Proceedings of International Symposium on Methodologies for Intelligent Systems*, pages 158–167. Springer.

[Bak, 1996] Bak, P. (1996). *How Nature Works: The Science of Self-Organized Criticality*. Copernicus Books.

[Balch, 1997] Balch, T. (1997). Learning roles: Behavioral diversity in robot teams. In *Proceedings of the AAAI-97 Workshop on Multi-Agent Learning*, pages 7–12.

[Balch and Arkin, 1995] Balch, T. and Arkin, R. C. (1995). Motor schema based formation control for multi-agent robot teams. In *Proceedings of the First International Conference on Multi-Agent Systems*, pages 10–16. AAAI Press.

[Baluja, 1994] Baluja, S. (1994). Population based incremental learning: A method for integrating genetic search based function optimization and competitive learning. Technical Report CMU-CS-94-163, School of Computer Science, Carnegie Mellon University.

[Baluja and Caruana, 1995] Baluja, S. and Caruana, R. (1995). Removing the genetics from the standard genetic algorithm. In Prieditis, A. and Russel, S., editors, *Proceedings of the Twelfth International Conference on Machine Learning (ICML'95)*, pages 38–46. Morgan Kaufmann Publishers.

[Barabasi and Albert, 1999] Barabasi, A. L. and Albert, R. (1999). Emergence of scaling in random networks. *Science*, 286:509–512.

[Barabasi et al., 2000] Barabasi, A. L., Albert, R., and Jeong, H. (2000). Scale-free characteristics of random networks: The topology of the World Wide Web. *Physica A*, 281:69–77.

[Barford et al., 1999] Barford, P., Bestavros, A., Bradley, A., and Crovella, M. (1999). Changes in Web client access patterns: Characteristics and caching implications. *World Wide Web*, 2:15–28.

[Barford and Crovella, 1998] Barford, P. and Crovella, M. (1998). Generating representative Web workloads for network and server performance evaluation. In Leutenegger, S., editor, *Proceedings of the ACM SIGMETRICS'98 Conference on Measurement and Modeling of Computer Systems*, pages 151–160. ACM Press.

[Barták, 1998] Barták, R. (1998). On-line guide to constraint programming. http://kti.mff.cuni.cz/~bartak/constraints/stochastic.html.

[Bednarz, 2000] Bednarz, S. W. (2000). Mission geography. http://geog.tamu.edu/sarah/humangeog/migration8.html.

[Bitner and Reingold, 1975] Bitner, J. R. and Reingold, E. M. (1975). Backtrack programming techniques. *Communications of the ACM*, 18(11):651–656.

[Bonabeau et al., 1999] Bonabeau, E., Dorigo, M., and Theraulaz, G. (1999). *Swarm Intelligence: From Natural to Artificial Systems*. Oxford University Press.

[Bonabeau et al., 2000] Bonabeau, E., Dorigo, M., and Theraulaz, G. (2000). Inspiration for optimization from social insect behavior. *Nature*, 406:39–42.

[Breslau et al., 1998] Breslau, L., Cao, P., Fan, L., Phillips, G., and Shenker, S. (1998). Web caching and Zipf-like distributions: Evidence and implications. Technical Report 1371, Computer Sciences Department, University of Wisconsin-Madison.

[Broder et al., 2000] Broder, A., Kumar, R., Maghoul, F., Raghavan, P., Rajagopalan, S., Stata, R., Tomkins, A., and Wiener, J. (2000). Graph structure in the Web. In Bulterman, D., editor, *Proceedings of the Ninth World Wide Web Conference (WWW9)*, pages 247–256.

[Brooks, 1991] Brooks, R. A. (1991). Intelligence without representation. *Artificial Intelligence*, 47:139–159.

[Bruynooghe, 1981] Bruynooghe, M. (1981). Solving combinatorial search problems by intelligent backtracking. *Information Processing Letters*, 12(91):36–39.

[Casti, 1997] Casti, J. (1997). *Would-Be Worlds: How Simulation is Changing the Frontiers of Science*. John Wiley & Son.

[Catledge and Pitkow, 1995] Catledge, L. D. and Pitkow, J. E. (1995). Characterizing browsing strategies in the World-Wide Web. *Computer Networks and ISDN Systems*, 26(6):1065–1073.

[Cbakrabarti et al., 1999] Cbakrabarti, S., Dom, B. E., Gibson, D., and Kleinberg, J. (1999). Mining the Web's link structure. *IEEE Computer*, 32(8):60–67.

[Chiusano et al., 1997] Chiusano, S., Corno, F., Prinetto, P., and Reorda, M. Sonza (1997). Cellular automata for sequential test pattern generation. In Nicholadis, M. and Singh, A., editors, *Proceedings of the Fifteenth IEEE VLSI Test Symposium*, pages 60–65. IEEE Press.

[Chung and Reynolds, 2000] Chung, C. J. and Reynolds, R. G. (2000). Knowledge based self-adaptation in evolutionary search. *International Journal of Pattern Recognition and Artificial Intelligence*, 14(1):19–33.

[Clark, 1986] Clark, W. A. V. (1986). *Human Migration*. Sage Publications.

[Coello, 2002] Coello, C. A. (2002). An updated survey of ga based multiobjective optimization techniques. *ACM Computing Surveys*, 32(2):109–143.

[Conte and Gilbert, 1995] Conte, R. and Gilbert, N. (1995). Introduction: Computer simulation for social theory. In Gilbert, N. and Conte, R., editors, *Artificial Societies*, chapter 1, pages 1–15. UCL Press.

[Cooley et al., 1997] Cooley, R., Srivastava, J., and Mobasher, B. (1997). Web mining: Information and pattern discovery on the World Wide Web. In *Proceedings of the Ninth IEEE International Conference on Tools with Artificial Intelligence (ICTAI'97)*, pages 558–567. IEEE Press.

[Cooley et al., 1999] Cooley, R., Tan, P. N., and Srivastava, J. (1999). Discovery of interesting usage patterns from Web data. In Masand, B. and Spiliopoulou, Myra, editors, *Proceedings of the Workshop on Web Usage Analysis and User Profiling (WEBKDD'99)*, pages 163–182. Springer.

[Cooper, 1989] Cooper, M. C. (1989). An optimal k-consistency algorithm. *Artificial Intelligence*, 41:89–95.

[Corno et al., 2000] Corno, F., Reorda, M. Sonza, and Squillero, G. (2000). Exploiting the selfish gene algorithm for evolving hardware cellular automata. In *Proceedings of the 2000 Congress on Evolutionary Computation (CEC2000)*, pages 1401–1406.

[Cove and Walsh, 1988] Cove, J. F. and Walsh, B.C. (1988). Online text retrieval via browsing. *Information Processing and Management*, 24(1):31–37.

[Crovella and Taqqu, 1999] Crovella, M. E. and Taqqu, M. S. (1999). Estimating the heavy tail index from scaling properties. *Methodology and Computing in Applied Probability*, 1(1):55–79.

[Crutchfield and Mitchell, 1995] Crutchfield, J. P. and Mitchell, M. (1995). The evolution of emergent computation. In *Proceedings of the National Academy of Sciences, USA*, volume 92, pages 10742–10746. National Academy of Sciences.

[Cuhna et al., 1995] Cuhna, C. R., Bestavros, A., and Crovella, M. E. (1995). Characteristics of WWW client based traces. Technical Report BU-CS-95-010, Computer Science Department, Boston University.

[Cuhna and Jaccoud, 1997] Cuhna, C. R. and Jaccoud, C. (1997). Determining WWW user's next access and its application to pre-fetching. In *Proceedings of the Second IEEE Symposium on Computers and Communications (ISCC'97)*, page 6. IEEE Press.

[Das et al., 1995] Das, R., Crutchfield, J. P., Mitchell, M., and Hanson, J. E. (1995). Evolving globally synchronized cellular automata. In Eshelman, L., editor, *Proceedings of the Sixth International Conference on Genetic Algorithms (ICGA-95)*, pages 336–343. Morgan Kaufmann Publishers.

[Davis et al., 1962] Davis, M., Logemann, G., and Loveland, D. (1962). A machine program for theorem proofing. *Communications of the ACM*, 5:394–397.

[DeJong, 1975] DeJong, K. A. (1975). *Analysis of the Behavior of a Class of Genetic Adaptive Systems*. PhD thesis, Department of Computer and Communication Sciences, University of Michigan.

[Doran and Gilbert, 1994] Doran, J. and Gilbert, N. (1994). Simulating societies: An introduction. In Gilbert, N. and Doran, J., editors, *Simulating Societies: The Computer Simulation of Social Phenomena*, chapter 1, pages 1–18. UCL Press.

[Dorigo et al., 1999] Dorigo, M., Caro, G. Di, and Gambardella, L. M. (1999). Ant algorithms for discrete optimization. *Artificial Life*, 5(2):137–172.

[Dorigo et al., 1991] Dorigo, M., Maniezzo, V., and Colorni, A. (1991). The ant system: An autocatalytic optimizing process. Technical Report 91-016 Revised, Politecnico di Milano.

[Dorigo et al., 1996] Dorigo, M., Maniezzo, V., and Colorni, A. (1996). The ant system: Optimization by a colony of cooperative agents. *IEEE Transactions on Systems, Man, and Cybernetics, Part B,* 26(1):1–13.

[Dunlap, 1997] Dunlap, R. A. (1997). *Golden Ratio and the Fibonnaci Numbers.* World Scientific.

[Durfee, 1999] Durfee, E. H. (1999). Distributed problem solving and planning. In Weiss, G., editor, *Multi-Agent Systems: A Modern Approach to Distributed Artificial Intelligence,* pages 121–164. MIT Press.

[Eschenauer et al., 1986] Eschenauer, H. A., Koski, J., and Osyczka, A. (1986). *Multicriteria Design Optimization: Procedures and Applications.* Springer.

[Ferber, 1999] Ferber, J. (1999). *Multi-agent Systems: An Introduction to Distributed Artificial Intelligence,* chapter 1, pages 31–35. Addison-Wesley.

[Flake et al., 2002] Flake, G. W., Lawrence, S., Giles, C. L., and Coetzee, F. (2002). Self-organization of the Web and identification of communities. *Computer,* 35(3):66–71.

[Fogel ct al., 1966] Fogel, L., Owens, A. J., and Walsh, M. J. (1966). *Artificial Intelligence Through Simulation Evolution.* John Wiley & Sons.

[Folino et al., 2001] Folino, G., Pizzuti, C., and Spezzano, G. (2001). Parallel hybrid method for SAT that couples genetic algorithms and local search. *IEEE Transactions on Evolutionary Computation,* 5(4):323–334.

[Freisleben, 1997] Freisleben, B. (1997). Metaevolutionary approaches. In T. Bäck, D. B. Fogel and Michalewicz, Z., editors, *Handbook of Evolutionary Computation,* chapter 7.2, pages 1–8. Oxford University Press.

[Freisleben and Härtfelder, 1993a] Freisleben, B. and Härtfelder, M. (1993a). In search of the best genetic algorithm for the traveling salesman problem. In *Proceedings of the Ninth International Conference on Control Systems and Computer Science (CSCS-9),* pages 485–493.

[Freisleben and Härtfelder, 1993b] Freisleben, B. and Härtfelder, M. (1993b). Optimization of genetic algorithms by genetic algorithms. In R. F. Albrecht, C. R. Reeves and Steele, N. C., editors, *Artificial Neural Networks and Genetic Algorithms,* pages 392–399. Springer.

[Freuder and Wallace, 1992] Freuder, E. C. and Wallace, R. J. (1992). Partial constraint satisfaction. *Artificial Intelligence,* 58(1-3):21–70.

[Gaschnig, 1979] Gaschnig, J. (1979). *Performance Measurement and Analysis of Certain Search Algorithm.* PhD thesis, Department of Computer Science, Carnegie-Mellon University.

[Geist et al., 1994] Geist, A., Beguelin, A., Dongarra, J., Jiang, W., Manchek, R., and Sunderam, V. (1994). *PVM: Parallel Virtual Machine, A Users' Guide and Tutorial for Networked Parallel Computing.* MIT Press.

[Gen et al., 2000] Gen, M., Zhou, G., and Kim, J. R. (2000). Genetic algorithms for solving network design problems: State-of-the-Art survey. *Evolutionary Optimization: An International Journal on the Internet,* 2(1):21–41.

[Gent and Walsh, 1993] Gent, I. P. and Walsh, T. (1993). Towards an understanding of hill-climbing procedures for SAT. In *Proceedings of the Eleventh National Conference on Artificial Intelligence (AAAI-93)*, pages 28–33. AAAI Press/MIT Press.

[Gent and Walsh, 1995] Gent, I. P. and Walsh, T. (1995). Unsatisfied variables in local search. In Hallam, J., editor, *Hybrid Problems, Hybrid Solutions*, pages 73–85. IOS Press.

[Gibson et al., 1998] Gibson, D., Kleinberg, J., and Raghavan, P. (1998). Inferring Web communities from link topology. In Akscyn, R., editor, *Proceedings of the Ninth ACM Conference on Hypertext and Hypermedia (HYPERTEXT'98)*, pages 225–234. ACM Press.

[Glassman, 1994] Glassman, S. (1994). A caching relay for the World Wide Web. *Computer Networks and ISDN Systems*, 27(2):165–173.

[Goldberg and Smith, 1987] Goldberg, D. E. and Smith, R. E. (1987). Nonstationary function optimization using genetic algorithms with dominance and diploidy. In Grefenstette, J., editor, *Proceedings of the Second International Conference on Genetic Algorithms (ICGA-87)*, pages 59–68.

[Goss et al., 1990] Goss, S., Beckers, R., Deneubourg, J. L., Aron, S., and Pasteels, J. M. (1990). How trail laying and trail following can solve foraging problems for ant colonies. In Hughes, R. N., editor, *Behavioral Mechanisms of Food Selection*, volume G20 of *NATO-ASI*, pages 661–678. Springer.

[Grefenstette, 1986] Grefenstette, J. J. (1986). Optimization of control parameters for genetic algorithms. *IEEE Transactions on Systems, Man and Cybernetics*, 16(1):122–128.

[Gu, 1992] Gu, J. (1992). Efficient local search for very large-scale satisfiability problem. *SIGART Bulletin*, 3:8–12.

[Gu, 1993] Gu, J. (1993). Local search for satisfiability (SAT) problem. *IEEE Transactions on Systems, Man, and Cybernetics*, 23(4):1108–1129.

[Gutowitz, 1991] Gutowitz, H. (1991). *Cellular Automata: Theory and Experiment*. MIT Press.

[GVU, 2001] GVU (2001). GVU's WWW user surveys. http://www.gvu.gatech.edu/user_surveys.

[Gzickman and Sycara, 1996] Gzickman, H. R. and Sycara, K. P. (1996). Self-adaptation of mutation rates and dynamic fitness. In *Proceedings of the Thirteenth National Conference on Artificial Intelligence (AAAI-96) and the Eighth Innovative Applications of Artificial Intelligence Conference on Artificial Intelligence (IAAI-96)*, volume 2, page 1389. MIT Press.

[Haken, 1983a] Haken, H. (1983a). *Advanced Synergetics : Instability Hierarchies of Self-Organizing Systems and Devices*. Springer.

[Haken, 1983b] Haken, H. (1983b). *Synergetics: An Introduction. Nonequilibrium Phase Transition and Self-Organization in Physics, Chemistry, and Biology*. Springer, third revised and enlarged edition edition.

[Haken, 1988] Haken, H. (1988). *Information and Self-Organization : A Macroscopic Approach to Complex Systems*. Springer.

[Han and Lee, 1988] Han, C. C. and Lee, C. H. (1988). Comments on Mohr and Henderson's path consistency algorithm. *Artificial Intelligence*, 36:125–130.

[Han et al., 1999] Han, J., Liu, J., and Cai, Q. S. (1999). From ALife agents to a kingdom of N queens. In Liu, J. and Zhong, N., editors, *Intelligent Agent Technology: Systems, Methodologies and Tools*, pages 110–120. World Scientific Publishing.

[Helbing et al., 2000a] Helbing, D., Farkas, I., and Vicsek, T. (2000a). Simulating dynamic features of escape panic. *Nature*, 407:487–490.

[Helbing and Huberman, 1998] Helbing, D. and Huberman, B. A. (1998). Coherent moving states in highway traffic. *Nature*, 396:738–740.

[Helbing et al., 2000b] Helbing, D., Huberman, B. A., and Maurer, S. M. (2000b). Optimizing traffic in virtual and real space. In Helbing, D., Herrmann, H. J., Schreckenberg, M., and Wolf, D. E., editors, *Traffic and Granular Flow '99: Social, Traffic, and Granular Dynamics*. Springer.

[Herrera and Lozano, 1998] Herrera, F. and Lozano, M. (1998). Adaptive genetic operators based on coevolution with fuzzy behaviors. Technical Report CMU-CS-94-163, Department of Computer Science and Artificial Intelligence, University of Granada.

[Hinterding et al., 1997] Hinterding, R., Michalewicz, Z., and Eiben, A. E. (1997). Adaptation in evolutionary computation: A survey. In Bäck, T., Michalewicz, Z., and Yao, X., editors, *Proceedings of the Fourth IEEE International Conference on Evolutionary Computation (ICEC'97)*, pages 65–69. IEEE Press.

[Hogg and Huberman, 1993] Hogg, T. and Huberman, B. A. (1993). Better than the best: The power of cooperation. In Nadel, L. and Stein, D., editors, *SFI 1992 Lectures in Complex Systems*, pages 163–184. Addison-Wesley.

[Holland, 1992] Holland, J. H. (1992). *Adaptation in Natural and Artificial Systems*. MIT Press.

[Hoos and Stützle, 1999] Hoos, H. H. and Stützle, T. (1999). Systematic vs. local search for SAT. In *Proceedings of KI-99*, volume 1701 of *LNAI*, pages 289–293. Springer.

[Hoos and Stützle, 2000a] Hoos, H. H. and Stützle, T. (2000a). Local search algorithms for SAT: An empirical evaluation. *Journal of Automated Reasoning*, 24:421–481.

[Hoos and Stützle, 2000b] Hoos, H. H. and Stützle, T. (2000b). SATLIB: An online resource for research on SAT. In Gent, I. P., Maaren, H. V., and Walsh, T., editors, *Proceedings of the Third Workshop on the Satisfiability Problem (SAT 2000)*, pages 283–292. IOS Press.

[Hordijk et al., 1998] Hordijk, W., Crutchfield, J. P., and Mitchell, M. (1998). Mechanisms of emergent computation in cellular automata. In Eiben, A. E., Bäck, T., Schoenauer, M., and Schwefel, H. P., editors, *PPSN V: Proceedings of the Fifth International Conference on Parallel Problem Solving from Nature*, volume 1498 of *LNCS*, pages 613–622. Springer.

[Horst and Pardalos, 1995] Horst, R. and Pardalos, P. M., editors (1995). *Handbook of Global Optimization*. Kluwer Academic Publishers.

[Horst and Tuy, 1990] Horst, R. and Tuy, H. (1990). *Global Optimization: Deterministic Approaches*. Springer.

[Howard, 1997] Howard, K. R. (1997). Unjamming traffic with computers. *Scientific American*, 277(4):86–88.

[Huberman, 1988] Huberman, B. A., editor (1988). *The Ecology of Computation*. North-Holland.

[Huberman and Adamic, 1999a] Huberman, B. A. and Adamic, L. A. (1999a). Evolutionary dynamics of the World Wide Web. *Nature*, 399:131.

[Huberman and Adamic, 1999b] Huberman, B. A. and Adamic, L. A. (1999b). Growth dynamics of the World Wide Web. *Nature*, 410:131.

[Huberman et al., 1997] Huberman, B. A., Pirolli, P. L. T., Pitkow, J. E., and Lukose, R. M. (1997). Strong regularities in World Wide Web surfing. *Science*, 280:96–97.

[IEEE, 2001] IEEE (2001). Draft standard for information technology, learning technology glossary. Technical Report P1484.3/D3, IEEE.

[IEEE, 2002] IEEE (2002). IEEE standard for learning object metadata. Technical Report P1484.12.1, IEEE.

[Ingber, 1996] Ingber, L. (1996). Adaptive simulated annealing (ASA): Lessons learned. *Journal of Control and Cybernetics*, 25:33–54.

[Jennings and Wooldridge, 1996] Jennings, N. R. and Wooldridge, M. (1996). Software agents. *IEE Review*, 42(1):17–21.

[Jensen, 1998] Jensen, H. J. (1998). *Self-Organized Criticality: Emergent Complex Behavior in Physical and Biological Systems*. Cambridge University Press.

[Johansen and Sornette, 2000] Johansen, A. and Sornette, D. (2000). Download relaxation dynamics on the WWW following newspapers publication of URL. *Physica A*, 276:338–345.

[Joshi and Krishnapuram, 2000] Joshi, A. and Krishnapuram, R. (2000). On mining Web access logs. In *Proceedings of the 2000 ACM SIGMOD Workshop on Research Issues in Data Mining and Knowledge Discovery*, pages 63–69.

[Kanada, 1992] Kanada, Y. (1992). Toward self-organization by computers. In *Proceedings of the Thirty-Third Programming Symposium, Information Processing Society of Japan*.

[Kanada and Hirokawa, 1994] Kanada, Y. and Hirokawa, M. (1994). Stochastic problem solving by local computation based on self-organization paradigm. In *Proceedings of the IEEE Twenty-Seventh Hawaii International Conference on System Sciences*, pages 82–91.

[Kauffman, 1993] Kauffman, S. (1993). *Origins of Order: Self-Organization and Selection in Evolution*. Oxford University Press.

[Kennedy, 1997] Kennedy, J. (1997). The particle Swarm: Social adaptation of knowledge. In *Proceedings of the Fourth IEEE International Conference on Evolutionary Computation (ICEC'97)*, pages 303–308.

[Kirkpatrick et al., 1983] Kirkpatrick, S., Gelatt, C. D., and Vecchi, M. P. (1983). Optimization by simulated annealing. *Science*, 220:671–680.

[Ko and Garcia, 1995] Ko, E. J. and Garcia, O. N. (1995). Adaptive control of crossover rate in genetic programming. In *Proceedings of the Artificial Neural Networks in Engineering (ANNIE'95)*. ASME Press.

[Ko et al., 1996] Ko, M. S., Kang, T. W., and Hwang, C. S. (1996). Adaptive crossover operator based on locality and convergence. In *Proceedings of 1996 IEEE International Joint Symposia on Intelligence and Systems (IJSIS'96)*, pages 18–22. IEEE Computer Society Press.

[Kuhnel, 1997] Kuhnel, R. (1997). Agent oriented programming with Java. In Plander, I., editor, *Proceedings of the Seventh International Conference on Artificial Intelligence and Information – Control Systems of Robots (AIICSR'97)*. World Scientific Publishing.

[Kumar, 1992] Kumar, V. (1992). Algorithm for constraint satisfaction problem: A survey. *AI Magazine*, 13(1):32–44.

[Langton, 1989] Langton, C. G. (1989). Artificial life. In Langton, C. G., editor, *Artificial Life*, volume VI of *SFI Studies in the Sciences of Complexity*, pages 1–47. Addison-Wesley.

[Langton, 1992] Langton, C. G. (1992). Preface. In Langton, C. G., Taylor, C., Farmer, J. D., and Rasmussen, S., editors, *Artificial Life II*, volume X of *SFI Studies in the Sciences of Complexity*, pages xiii–xviii. Addison-Wesley.

[Lawrence and Giles, 1999] Lawrence, S. and Giles, C. L. (1999). Accessibility of information on the Web. *Nature*, 400:107–109.

[Lee and Takagi, 1993] Lee, M. A. and Takagi, H. (1993). Dynamic control of genetic algorithms using fuzzy logic techniques. In Forrest, S., editor, *Proceedings of the Fifth International Conference on Genetic Algorithms (ICGA-93)*, pages 76–83.

[Levene et al., 2001] Levene, M., Borges, J., and Loizou, G. (2001). Zipf's law for Web surfers. *Knowledge and Information Systems*, 3:120–129.

[Levene and Loizou, 1999] Levene, M. and Loizou, G. (1999). Computing the entropy of user navigation in the Web. Technical Report RN/99/42, Department of Computer Science, University College London.

[Liang et al., 1998] Liang, K. H., Yao, X., and Newton, C. (1998). Dynamic control of adaptive parameters in evolutionary programming. In *Proceedings of the Second Asia-Pacific Conference on Simulated Evolution and Learning (SEAL'98)*, pages 42–49. Springer.

[Liu, 2001] Liu, J. (2001). *Autonomous Agents and Multi-Agent Systems: Explorations in Learning, Self-Organization, and Adaptive Computation*. World Scientific Publishing.

[Liu and Han, 2001] Liu, J. and Han, J. (2001). ALIFE: A multi-agent computing paradigm for constraint satisfaction problems. *International Journal of Pattern Recognition and Artificial Intelligence*, 15(3):475–491.

[Liu et al., 2002] Liu, J., Han, J., and Tang, Y. Y. (2002). Multi-agent oriented constraint satisfaction. *Artificial Intelligence*, 136(1):101–144.

[Liu et al., 2004a] Liu, J., Jin, X., and Tsui, K. C. (2004a). Autonomy oriented computing (AOC): Formulating computational systems with autonomous components. *IEEE Transactions on Systems, Man and Cybernetics, Part A: Systems and Humans (in press)*.

[Liu and Tang, 1999] Liu, J. and Tang, Y. Y. (1999). Adaptive image segmentation with distributed behavior based agents. *IEEE Transactions on Pattern Analysis and Machine Intelligence*, 21(6):544–551.

[Liu et al., 1997] Liu, J., Tang, Y. Y., and Cao, Y. C. (1997). An evolutionary autonomous agents approach to image feature extraction. *IEEE Transactions on Evolutionary Computation*, 1(2):141–158.

[Liu and Tsui, 2001] Liu, J. and Tsui, K. C. (2001). Introducing autonomy oriented computation. In *Proceedings of the First International Workshop on Autonomy Oriented Computation (AOC'01)*, pages 1–11.

[Liu and Wu, 2001] Liu, J. and Wu, J. (2001). *Multi-Agent Robotic Systems*. CRC Press.

[Liu et al., 2004b] Liu, J., Zhang, S., and Yang, J. (2004b). Characterizing Web usage regularities with information foraging agents. *IEEE Transactions on Knowledge and Data Engineering*, 16(5):566–584.

[Loser et al., 2002] Loser, A., Grune, C., and Hoffmann, M. (2002). A didactic model, definition of learning objects and selection of metadata for an online curriculum. http://www.ibi.tu-berlin.de/diskurs/onlineduca/onleduc02/ Talk_Online_Educa__02_Loeser_TU_berlin.pdf.

[Louzoun et al., 2000] Louzoun, Y., Solomon, S., Atlan, H., and Cohen, I. R. (2000). The emergence of spatial complexity in the immune system. Los Alamos Physics Archive arXiv:cond-mat/0008133, http://xxx.lanl.gov/html/cond-mat/0008133.

[Lucas, 1997] Lucas, C. (1997). Self-organizing systems (SOS). http://www.calresco.org/ sos/sosfaq.htm.

[Lukose and Huberman, 1998] Lukose, R. M. and Huberman, B. A. (1998). Surfing as a real option. In *Proceedings of the First International Conference on Information and Computation Economics (ICE'98)*, pages 45–51.

[Mackworth, 1977] Mackworth, A. K. (1977). Consistency in networks of relations. *Artificial Intelligence*, 8(1):99–118.

[Madria et al., 1999] Madria, S., Bhowmick, S. S., NG, W. K., and Lim, R. P. (1999). Research issues in Web data mining. In *Proceedings of the First International Conference on Data Warehousing and Knowledge Discovery (DAWAK99)*, volume 1676 of *LNCS*, pages 303–312. Springer.

[Mataric, 1994] Mataric, M. J. (1994). Reward functions for accelerated learning. In Cohen, W. W. and Hirsh, H., editors, *Proceedings of the Eleventh International Conference on Machine Learning (ICML'94)*, pages 181–189. Morgan Kaufmann Publishers.

[Maurer and Huberman, 2000] Maurer, S. M. and Huberman, B. A. (2000). The competitive dynamics of Web sites. http://ideas.repec.org/p/sce/scecf0/357.html.

[Mazure et al., 1997] Mazure, B., Sais, L., and Grégoire, É. (1997). Tabu search for SAT. In *Proceedings of the Fourteenth National Conference on Artificial Intelligence (AAAI'97)*, pages 281–285.

[McAllester et al., 1997] McAllester, D., Selman, B., and Kautz, H. (1997). Evidence for invariants in local search. In *Proceedings of the Fourteenth National Conference on Artificial Intelligence (AAAI'97)*, pages 321–326.

[Menczer, 2004a] Menczer, F. (2004a). Lexical and semantic clustering by Web links. *Journal of the American Society for Information Science and Technology (in press)*.

[Menczer, 2004b] Menczer, F. (2004b). Mapping the semantics of Web text and links. Working paper, http://www.informatics.indiana.edu/fil/papers.asp.

[Michalewicz, 1994] Michalewicz, Z. (1994). *Genetic Algorithms + Data Structures = Evolution Programs*. Springer.

[Milgram, 1967] Milgram, S. (1967). The small world problem. *Psychology Today*, 2:60–67.

[Minar et al., 1996] Minar, N., Burkhart, R., Langton, C. G., and Askenazi, M. (1996). The Swarm simulation system: A toolkit for building multi-agent simulations. Santa Fe Institute, http://www.santafe.edu/projects/swarm/overview/overview.html.

[Minton et al., 1992] Minton, S., Johnston, M. D., Philips, A., and Laird, P. (1992). Minimizing conflicts: A heuristic repair method for constraint satisfaction and scheduling problems. *Artificial Intelligence*, 58:161–205.

[Mobasher et al., 1996] Mobasher, B., Jain, N., Han, E., and Srivastava, J. (1996). Web mining: Pattern discovery from World Wide Web transactions. Technical Report TR-96050, Department of Computer Science, University of Minnesota.

[Mockus, 1989] Mockus, J. (1989). *Bayesian Approach to Global Optimization : Theory and Applications*. Kluwer Academic Publishers.

[Mogul, 1995] Mogul, J. (1995). Network behavior of a busy Web server and its clients. Technical Report TR-95.5, Digital Western Research Laboratory.

[Mohr and Henderson, 1986] Mohr, R. and Henderson, T. C. (1986). Arc and path consistency revisited. *Artificial Intelligence*, 28:225–233.

[Montoya and Sole, 2000] Montoya, J. M. and Sole, R. V. (2000). Small world patterns in food webs. http://arxiv.org/abs/cond-mat/0011195.

[MPI, 1996] MPI (1996). http://www-unix.mcs.anl.gov/mpi/.

[Müller et al., 2002] Müller, S. D., Marchetto, J., Airaghi, S., and Koumoustsakos, P. (2002). Optimization based on bacterial chemotaxis. *IEEE Transactions on Evolutionary Computation*, 6(1):16–29.

[Nadel, 1990] Nadel, B. (1990). Some applications of the constraint satisfaction problem. Technical Report CSC-90-008, Computer Science Department, Wayne State University.

[Nasraoui et al., 1999] Nasraoui, O., Frigui, H., Joshi, A., and Krishnapuram, R. (1999). Mining Web access logs using relational competitive fuzzy clustering. In *Proceedings of the Eighth International Fuzzy Systems Association World Congress (IFSA'99)*.

[Nehaniv, 2000a] Nehaniv, C., editor (2000a). *Proceedings of the Evolvability Workshop at the Seventh International Conference on the Simulation and Synthesis of Living Systems (Artificial Life 7)*. Published as University of Hertfordshire Technical Report 351.

[Nehaniv, 2000b] Nehaniv, C. L. (2000b). Measuring evolvability as the rate of complexity increase. In *Nehaniv [Nehaniv, 2000a]*, pages 66–68.

[Nehaniv, 2000c] Nehaniv, C. L. (2000c). Preface. In *Nehaniv [Nehaniv, 2000a]*, pages iii–iv.

[Nehaniv and Rhodes, 2000] Nehaniv, C. L. and Rhodes, J. L. (2000). The evolution and understanding of hierarchical complexity in biology from an algebraic perspective. *Artificial Life*, 6(1):45–67.

[Nicolis and Prigogine, 1977] Nicolis, G. and Prigogine, I. (1977). *Self-Organization in Non-Equilibrium Systems: From Dissipative Structures to Order through Fluctuations.* John Wiley.

[Nwana et al., 1998] Nwana, H. S., Ndumu, D. T., and Lee, L. C. (1998). ZEUS: An advanced tool-kit for engineering distributed multi-agent systems. In *Proceedings of the Third International Conference on the Practical Applications of Intelligent (PAAM'98)*, pages 377–391.

[Padmanabhan and Mogul, 1996] Padmanabhan, V. and Mogul, J. (1996). Using predictive prefetching to improve World Wide Web latency. In *Proceedings of the ACM SIGCOMM Conference on Applications, Technologies, Architectures and Protocols for Computer Communication (SIGCOMM'96)*, pages 22–36.

[Pavlidis, 1992] Pavlidis, T. (1992). *Algorithms for Graphics and Image Processing*. Computer Science Press.

[Pei et al., 2000] Pei, J., Han, J., Mortazavi-asl, B., and Zhu, H. (2000). Mining access patterns efficiently from Web logs. In *Proceedings of the Pacific-Asia Conference on Knowledge Discovery and Data Mining (PAKDD2000)*, pages 396–407.

[Perold, 1984] Perold, A. F. (1984). Large-scale portfolio optimization. *Management Science*, 30:1143–1160.

[Pitas, 1993] Pitas, I. (1993). *Digital Image Processing Algorithms*. Prentice Hall.

[Pitkow, 1998] Pitkow, J. E. (1998). Summary of WWW characterizations. *Computer Networks and ISDN Systems*, 30:551–558.

[Prigogine, 1980] Prigogine, I. (1980). *From Being to Becoming: Time and Complexity in the Physical Sciences*. W. H. Freeman and Company.

[Prusinkiewicz et al., 1997] Prusinkiewicz, P., Hammel, M., and Mech, R. (1997). Visual models of morphogenesis: A guided tour. http://algorithmicbotany.org/vmm/title.html.

[Prusinkiewicz and Lindenmayer, 1990] Prusinkiewicz, P. and Lindenmayer, A. (1990). *The Algorithmic Beauty of Plants*. Springer.

[PVM, 1989] PVM (1989). http://www.epm.ornl.gov/pvm/.

[Rasmussen and Barrett, 1995] Rasmussen, S. and Barrett, C. (1995). Elements of a theory of simulation. Technical Report 95-04-040, Santa Fe Institute.

[Ray, 1992] Ray, T. S. (1992). An approach to the synthesis of life. In Langton, C. G., Taylor, C., Farmer, J. D., and Rasmussen, S., editors, *Artificial Life II*, volume X of *SFI Studies in the Sciences of Complexity*, pages 371–408. Addison-Wesley.

[Resnick, 1994] Resnick, M. (1994). *Turtles, Termites and Traffic Jams: Explorations in Massively Parallel Microworlds*. MIT Press.

[Reynolds, 1994] Reynolds, R. G. (1994). An introduction to cultural algorithms. In Sebald, A. V. and Fogel, L. J., editors, *Proceedings of the Third Annual Conference on Evolutionary Programming (EP'94)*, pages 131–139.

[Ronald et al., 1999] Ronald, E. M. A., Sipper, M., and Capcarrère, M. S. (1999). Design, observation, surprise! A test of emergence. *Artificial Life*, 5(3):225–239.

[Rossi et al., 1990] Rossi, F., Petrie, C., and Dhar, V. (1990). On the equivalence of constraint satisfaction problem. In *Proceedings of the Ninth European Conference on Artificial Intelligence (ECAI-90)*, pages 550–556.

[Sandholm, 1999] Sandholm, T. W. (1999). Distributed rational decision making. In Weiss, G., editor, *Multi-Agent Systems: A Modern Approach to Distributed Artificial Intelligence*, pages 201–258. MIT Press.

[SATLIB, 2000] SATLIB (2000). http://www.intellektik.informatik.tu-darmstadt.de/SATLIB/.

[Schwefel, 1981] Schwefel, H. P. (1981). *Numerical Optimization of Computer Models*. John Wiley & Sons.

[Schwefel, 1995] Schwefel, H. P. (1995). *Evolution and Optimum Seeking*. John Wiley & Sons.

[Sebag and Schoenauer, 1996] Sebag, M. and Schoenauer, M. (1996). Mutation by imitation in Boolean evolution strategies. In Voigt, H. M., Ebeling, W., Rechenberg, I., and Schwefel, H.-P., editors, *PPSN IV: Proceedings of the Fourth Conference on Parallel Problem Solving from Nature*, volume 1141 of *LNCS*, pages 356–365. Springer.

[Selman et al., 1994] Selman, B., Kautz, H., and Cohen, B. (1994). Noise strategies for improving local search. In *Proceedings of the Twelfth National Conference on Artificial Intelligence (AAAI'94)*, pages 337–343.

[Selman et al., 1992] Selman, B., Levesque, H., and Mitchell, D. (1992). A new method of solving local search. In *Proceedings of the Ninth National Conference on Artificial Intelligence (AAAI'92)*, pages 440–446.

[Shanahan, 1994] Shanahan, M. (1994). Evolutionary automata. In *Artificial Life IV: Proceedings of the Fourth International Workshop Synthesis and Simulation of Living Systems*, pages 387–393. MIT Press.

[Shoham, 1993] Shoham, Y. (1993). Agent oriented programming. *Artificial Intelligence*, 60(1):51–92.

[Silaghi et al., 2001a] Silaghi, M., Haroud, D., and Faltings, B. (2001a). ABT with asynchronous reordering. In *Proceedings of the International Conference on Intelligent Agent Technology (IAT'01)*, pages 54–63.

[Silaghi et al., 2001b] Silaghi, M., Haroud, D., and Faltings, B. (2001b). Asynchronous consistency maintenance. In *Proceedings of the International Conference on Intelligent Agent Technology (IAT'01)*, pages 98–102.

[Silaghi et al., 2001c] Silaghi, M., Haroud, D., and Faltings, B. (2001c). Secure asynchronous search. In *Proceedings of the International Conference on Intelligent Agent Technology (IAT'01)*, pages 400–404.

[Sims, 1991] Sims, K. (1991). Artificial evolution for computer graphics. *Computer Graphics*, 25(4):319–328.

[Smith and Fogarty, 1996] Smith, J. E. and Fogarty, T. C. (1996). Adaptive parameterized evolutionary systems: Self adaptive recombination and mutation in genetic algorithm. In Voigt, H. M., Ebeling, W., Rechenberg, I., and Schwefel, H. P., editors, *PPSN IV: Proceedings of the Fourth Conference on Parallel Problem Solving from Nature*, volume 1141 of *LNCS*, pages 441–450. Springer.

[Smith and Taylor, 1998] Smith, R. E. and Taylor, N. (1998). A framework for evolutionary computation in agent based systems. In *Proceedings of the 1998 International Conference on Intelligent Systems*, pages 221–224. ISCA Press.

[Snir et al., 1996] Snir, M., Otto, S., Huss-Lederman, S., Walker, D., and Dongarra, J. (1996). *MPI: The Complete Reference*. MIT Press.

[Sosic and Gu, 1994] Sosic, R. and Gu, J. (1994). Efficient local search with conflict minimization: A case study of the n-queen problem. *IEEE Transactions on Knowledge and Data Engineering*, 6(5):661–668.

[Spiliopoulou, 1999] Spiliopoulou, M. (1999). The laborious way from data mining to Web log mining. *International Journal of Computer Systems Science and Engineering: Special Issue on Semantics of the Web*, 14:113–126.

[Spiliopoulou et al., 1999] Spiliopoulou, M., Pohle, C., and Faulstich, L. (1999). Improving the effectiveness of a Web site with Web usage mining. In *Proceedings of the Workshop on Web Usage Analysis and User Profiling (WEBKDD'99)*, pages 51–56. Springer.

[Stallman and Sussman, 1977] Stallman, R. and Sussman, G. J. (1977). Forward reasoning and dependency directed backtracking. *Artificial Intelligence*, 9(2):135–196.

[Standish, 1999] Standish, R. K. (1999). Some techniques for the measurement of complexity in Tierra. In Floreano, D., Nicoud, J. D., and Mondada, F., editors, *Advances in Artificial Life: The Proceeding of the Fifth European Conference on Artificial Life (ECAL'99)*, pages 104–108. Springer.

[Standish, 2001] Standish, R. K. (2001). On complexity and emergence. Los Alamos Physics Archive arXiv:nlin.AO/0101006, http://xxx.lanl.gov/abs/nlin/0101006.

[StarLogo, 2000] StarLogo (2000). http://www.media.mit.edu/starlogo/.

[Steinmann et al., 1997] Steinmann, O., Strohmaier, A., and Stützle, T. (1997). Tabu search vs. random walk. In *Advances in Artificial Intelligence (KI97)*, volume 1303 of *LNCS*, pages 337–348. Springer.

[Still, 2000] Still, G. K. (2000). *Crowd Dynamics*. PhD thesis, Mathematics Department, Warwick University.

[Storn and Price, 1997] Storn, R. and Price, K. (1997). Differential evolution – a simple and efficient adaptive scheme for global optimization over continuous spaces. *Journal of Global Optimization*, 11(4):341–359.

[Swain and Morris, 2000] Swain, A. K. and Morris, A. S. (2000). A novel hybrid evolutionary programming method for function optimization. In *Proceedings of the 2000 Congress on Evolutionary Computation (CEC2000)*, pages 1369–1376.

[Swarm, 1994] Swarm (1994).   An overview of the Swarm simulation system. http://www.santafe.edu/projects/swarm/swarm-blurb/swarm-blurb.html.

[Tang et al., 2003] Tang, Y., Liu, J., and Jin, X. (2003). Adaptive compromises in distributed problem solving. In *Proceedings of the Fourth International Conference on Intelligent Data Engineering and Automated Learning (IDEAL 2003)*, volume 2690 of *LNCS*, pages 35–42. Springer.

[Tettamanzi, 1995] Tettamanzi, A. G. (1995). Evolutionary algorithms and fuzzy logic: A two-way integration. In *Proceedings of the Second Joint Conference on Information Sciences (JCIS-95)*, pages 464–467.

[Thatcher, 1999] Thatcher, A. (1999). Determining interests and motives in WWW navigation. In *Proceedings of the Second International Cyberspace Conference on Ergonomics (CybErg1999)*.

[Torn and Zilinskas, 1989] Torn, A. and Zilinskas, A. (1989). *Global Optimization*. Springer.

[TSP, 2002] TSP (2002). http://www.math.princeton.edu/tsp/.

[Ünsal, 1993] Ünsal, C. (1993). Self-organization in large populations of mobile robots. Master's thesis, Department of Electrical Engineering, Virginia Polytechnic Institute and State University. http://www-2.cs.cmu.edu/~unsal/thesis/cemsthesis.html.

[Wallace, 1996] Wallace, R. (1996). Analysis of heuristic methods for partial constraint satisfaction problem. In *Principles and Practice of Constraint Programming (CP-1996)*, pages 482–496.

[Walsh, 1999] Walsh, T. (1999). Search in a small world. In *Proceedings of the Sixteenth International Joint Conference on Artificial Intelligence (IJCAI'99)*, pages 1172–1177.

[Watts and Strogatz, 1998] Watts, D. J. and Strogatz, S. H. (1998). Collective dynamics of small world networks. *Nature*, 393:440–442.

[Williams and Crossley, 1997] Williams, E. A. and Crossley, W. A. (1997). Empirically derived population size and mutation rate guidelines for a genetic algorithm with uniform crossover. In *Proceedings of the Second On-line World Conference on Soft Computing in Engineering Design and Manufacturing WSC2*, pages 163–172. Springer.

[Wright et al., 2000] Wright, W. A., Smith, R. E., Danek, M., and Greenway, P. (2000). A measure of emergence in an adapting, multi-agent context. In *Proceedings Supplement of SAB'2000*, pages 20–27.

[Yan et al., 1996] Yan, T. W., Jacobsen, M., Garcia-Molina, H., and Dayal, U. (1996). From user access patterns to dynamic hypertext linking. In *Proceedings of the Fifth World Wide Web Conference (WWW5)*, pages 1007–1014.

[Yao and Liu, 1997] Yao, X. and Liu, Y. (1997). Fast evolution strategies. *Control and Cybernetics*, 26(3):467–496.

[Yao et al., 1999] Yao, X., Liu, Y., and Lin, G. (1999). Evolutionary programming made faster. *IEEE Transaction on Evolutionary Computation*, 3(2):82–102.

[Yokoo, 1995] Yokoo, M. (1995). Asynchronous weak-commitment search for solving large-scale distributed CSPs. In *Proceedings of the First International Conference on Multi-Agent Systems (ICMAS'95)*, pages 467–518.

[Yokoo et al., 1998] Yokoo, M., Durfee, E. H., Ishida, T., and Kuwabara, K. (1998). The distributed constraint satisfaction problem: Formalization and algorithms. *IEEE Transactions on Knowledge and Data Engineering*, 10(5):673–685.

[Yokoo et al., 2001] Yokoo, M., Etzioni, O., Ishida, T., Jennings, N., and Sycara, K. (2001). *Distributed Constraint Satisfaction Foundations of Cooperation in Multi-Agent Systems*. Springer.

[Yokoo and Hirayama, 1998] Yokoo, M. and Hirayama, K. (1998). Distributed constraint satisfaction algorithm for complex local problems. In *Proceedings of the Third International Conference on Multi-Agent Systems (ICMAS'98)*, pages 372–379.

[Yokoo and Hirayama, 2000] Yokoo, M. and Hirayama, K. (2000). Algorithms for distributed constraint satisfaction: A review. *Autonomous Agents and Multi-Agent Systems*, 3(2):185–207.

[Yokoo and Kitamura, 1996] Yokoo, M. and Kitamura, Y. (1996). Multi-agent real-time-A* with selection: Introducing competition in cooperative search. In *Proceedings of the Second International Conference on Multi-Agent Systems (ICMAS'96)*, pages 409–416.

[Zaane et al., 1998] Zaane, O. R., Xin, M., and Han, J. (1998). Discovering Web access patterns and trends by applying OLAP and data mining technology on Web logs. In *Proceedings of the Fourth Annual Advances in Digital Libraries Conference (ADL'98)*, pages 19–29.

[Zhang and Shimohara, 2000] Zhang, Y. and Shimohara, K. (2000). A note on evolvability in Tierra. In *Nehaniv [Nehaniv, 2000a]*, pages 62–65.

[Zipf, 1949] Zipf, G. K. (1949). *Human Behavior and the Principle of Least Effort*. Addison-Wesley.

Index